Programmable Logic Controller (PLC) Tutorial

Circuits and programs for Rockwell Allen-Bradley MicroLogix and SLC 500 programmable controllers

for Electrical Engineers and Technicians

Stephen P. Tubbs, P.E.
formerly of the
Pennsylvania State University,
currently
working in industry

NOTICE TO THE READER

The author does not warrant or guarantee any of the products described herein. Also, the author does not warrant or guarantee any of the equipment or programs described herein or accept liability for any damages resulting from their use.

The reader is warned that electricity and the construction of electrical equipment are dangerous. It is the responsibility of the reader to use common sense and safe electrical and mechanical practices.

MicroLogix, RSLinx, RSLogix, and SLC 500 are trademarks of Allen-Bradley a division of Rockwell Automation.

Printed in the United States of America

ISBN   0-9659446-6-2

CONTENTS

INTRODUCTION

The purpose of this book is to teach and demonstrate the basics of the Allen-Bradley MicroLogix 1000 programmable logic controller. Information is provided to help the reader get and operate an inexpensive MicroLogix 1000 and associated hardware and software. Examples with ladder diagrams and circuit diagrams are provided to demonstrate different MicroLogix 1000 capabilities. Background information is also provided to relate the MicroLogix 1000 to other programmable logic controllers.

To most people a programmable controller, programmable logic controller, and PLC are the same thing. It should be noted, however, that PLC is a registered trademark that Allen-Bradley, a division of Rockwell Automation, uses to describe one of its lines of programmable controllers.

Programmable controllers are simply special purpose computers that control electrically operated processes. The processes might be in chemical plants, steel mills, or other types of industrial plants that need precise control. Often programmable controllers do jobs that were formerly done by networks of relays and/or by teams of human operators.

Programmable controllers are more flexible, more rugged, easier to reprogram, and less expensive than all but the simplest relay logic systems. Also, programmable controllers can receive more types of inputs than relays. For example, some can receive input data from bar code scanners. Programmable controller knowledge is useful for maintenance technicians, plant engineering personnel, technologists, and engineers.

Any process controlled by a programmable controller has the following:
1. The process being controlled.
2. Input devices such as switches, sensors, or pushbuttons.
3. Input modules that act as a protective boundary and convert signals from the input devices to a form that can be used by the programmable controller's central processing unit, communication, and memory.
4. The programmable controller's central processing unit (CPU), communication and memory, and power supply.
5. A software program.
6. Output modules that act as a protective boundary and convert the CPU's output to a form that can operate external devices.
7. External devices such as lights, solenoids, and motor starters.
8. An operator terminal for programming and monitoring the control system and process.

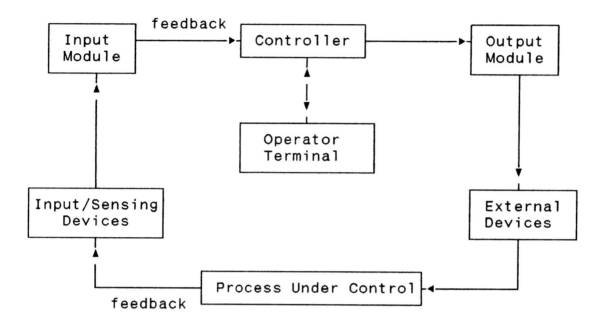

Figure I-1 Typical programmable controller configuration.

This book focuses on the relatively inexpensive 'student' version of the MicroLogix 1000 Allen-Bradley PLC, the RSLogix 500 PLC programming software, and the RSLinx software that interfaces the MicroLogix 1000 with a personal computer. The software and needed PLC wiring techniques learned with the 'student' version MicroLogix 1000 PLC are directly applicable to other MicroLogix PLCs and Allen-Bradley SLC 500 PLCs and are also useful with larger Allen-Bradley PLCs. Unlike many other PLCs, the MicroLogix and SLC 500 PLCs use the decimal system for numbering inputs and outputs. Other numbering systems, like binary, octal, hexadecimal, and binary coded decimal, will not be covered in this book.

The person using this book should already have a basic electrical education. He should be able to read electrical circuit diagrams. Roughly, he should be educated to or beyond the level of the graduate of a two-year electrical technician's program.

The reader should be ready to contact Allen-Bradley personnel for up-to-date help. Although efforts have been made to make this book up-to-date and accurate, the ever-increasing number of options and modifications available with the MicroLogix 1000 and RSLogix 500 software may make some of this book's information dated.

Stephen P. Tubbs

1.0 MICROLOGIX AND SLC 500 FAMILIES OF PLCS

MicroLogix and SLC 500 PLCs are the smallest, simplest, and least expensive PLCs made by Allen-Bradley (part of Rockwell Automation). They share the same architecture, instruction set and RSLogix 500 programming software. It isn't necessary to completely reprogram or learn a new system when moving from level to level within the MicroLogix and SLC 500 families.

The MicroLogix PLCs are physically smaller than SLC 500 PLCs. The smallest MicroLogix PLCs (the 1000 Series) are totally housed in small boxes (as small as 4.7" x 3.15" x 1.57"). Larger MicroLogix PLCs have their controllers and some inputs and outputs in a main box and may also have some terminals on plug-in modules. SLC 500s are chassis-based modular PLCs, physically similar to Allen-Bradley's older PLC-5. SLC 500s have their processors and all inputs and outputs on modules that are plugged into an SLC chassis.

The MicroLogix comes in three series, the 1000, 1200, and 1500. The 1200 and 1500 have more features and accept plug-in modules.

The MicroLogix 1000 Series family has the smallest and simplest PLCs. However, even with the MicroLogix 1000 it is possible to get up to 32 I/O (input and output terminals) with AC, DC, analog current, or analog voltage inputs and with relay, TRIAC, MOSFET, analog current, or analog voltage outputs. As with other PLCs, programming can be done with a personal computer or a special programming terminal.

2.0 DOWNLOADING PROGRAMMING SOFTWARE

The MicroLogix 1000 can be programmed, monitored, and troubleshot by a hand-held programmer (HHP) or by a personal computer (PC).

The hand-held-programmer is relatively inexpensive, light, and easy to connect. However, its display is more limited than that of a personal computer. The MicroLogix 1000 and most other Allen-Bradley programmable controllers are programmed with ladder logic. In ladder logic programming the program statements are entered on rungs that look like ladder rungs. With the hand-held programmer only one rung is shown at a time, making the program more difficult to follow. Also, the hand-held programmer does not have a full keyboard, making program entry more awkward. Hand-held-programmers can be purchased from electrical supply companies and occasionally on eBay. This book does not describe hand-held-programmer programming methods.

A personal computer, communications cable, and Allen-Bradley RSLinx and RSLogix software may be used instead of a hand-held-programmer. A personal computer as a programmer has the drawback of requiring greater setup time and being larger, heavier, and less rugged. However, the computer screen will show multiple rungs and is generally easier to use.

RSLinx and RSLogix software require the following hardware:
Pentium II processor with at least 128 MB of RAM
80 MB of available hard drive space
256 color, SVGA graphics adapter with 800 by 600 or greater resolution
CD ROM drive
Windows compatible mouse
Ethernet-card and/or Allen-Bradley communications device or cable

RSLinx and RSLogix software require one of the operating systems:
Microsoft Windows XP
Microsoft Windows 2000
Microsoft Windows Me
Microsoft Windows 98
Microsoft Windows NT, Version 4.0 with Service pack 6 or greater.

The software will not run under Windows 3.1, Windows for Workgroups using the 32-bit extensions or Windows NT 3.51.

4

2.1 TO GET FREE RSLOGIX 500 SOFTWARE

RSLogix 500 software is used to write ladder logic programs on a personal computer.

2.1.1 Go to the web address http://www.ab.com/plclogic/micrologix/

2.1.2 Look under 'GET SOFTWARE' and select 'RSLogix 500 Starter Software for 10 I/O (10-point) MicroLogix 1000 Controllers'.

2.1.3 A popup titled, 'Download Free Version of RSLogix 500 Starter Software' will appear.

2.1.4 Click on 'Take me to Downloads'.

2.1.5 A page will come up titled, 'Rockwell Automation Online Registration System'.

2.1.6 Register and/or login.

2.1.7 This should take you to the web page http://support.rockwellautomation.com/webupdates/

2.1.8 On this page there are a number of products available for downloading.

2.1.9 Select RSLogix 500 and click SUBMIT.

2.1.10 This will send you to a web page titled 'Software Updates'.

2.1.11 Select the software titled, 'RSLogix 500 Starter for 10 Point MicroLogix 1000 Programming Software'. It is numbered '9324RL0050ENE'. The latest version at the time this is being written is '6.10.00'. It is a web release.

2.1.12 Once the selection has been made you will be referred to a page that asks for your name.

2.1.13 After that you will be sent to a page that asks which files you want downloaded and how you want them downloaded.

2.1.14 RSLogix 500 is a big download. Depending on the speed of your connection you may want to download parts at a time. With a slow modem it can take many hours to download the needed files. If there is a loss of connection in the middle of downloading and you selected to download all files at once, you would have to schedule another long session to try again. If you download the files separately it may be easier to do the downloading.

2.1.15 Run the setup program to rejoin and install the RSLogix 500 files.

2.1.16 RSLogix 500 should now be ready for use.

2.2 TO GET FREE RSLINX LITE SOFTWARE

RSLinx Lite software is used to transfer RSLogix 500 ladder diagram programs to a 10-point MicroLogix 1000.

2.2.1 Repeat steps 2.1.1 to 2.1.7 used to get RSLogix 500 software.

2.2.2 Under "Communications Products' select RSLinx and click SUBMIT.

2.2.3 Select the software titled, 'RSLinx Lite Software'. It is numbered '9355WAB100ENE'. The latest version at the time this is being written is '2.43.01'. It is a general release.

2.2.4 Repeat steps 2.1.12 to 2.1.16 with the RSLinx Lite software.

3.0 MICROLOGIX 1000 TEACHING SETUP

A relatively inexpensive MicroLogix 1000 teaching setup can be built by a person with moderate soldering, wiring, and carpentry skills. As the photo below shows, the power supply, MicroLogix 1000, and input switch box were all mounted on a white pine board. Output devices were not installed, the MicroLogix 1000's output LEDs indicated the outputs' statuses.

The reader may wish to modify the construction to better fit his available equipment. The equipment photos and bill of materials can be used as a guide rather than exact plans.

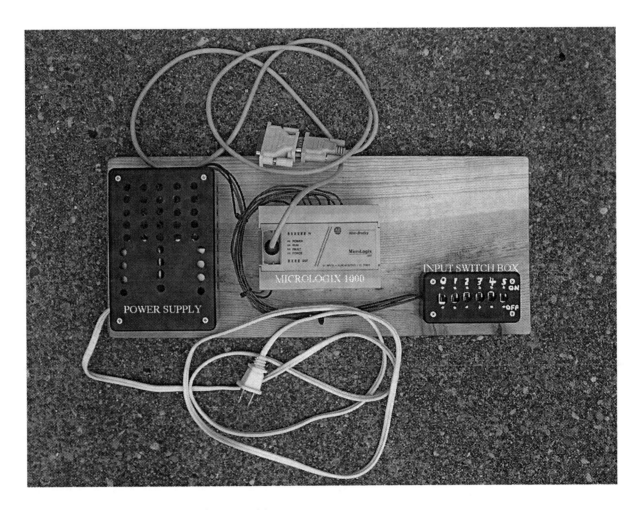

Figure 3-1 Photo of a complete teaching setup.

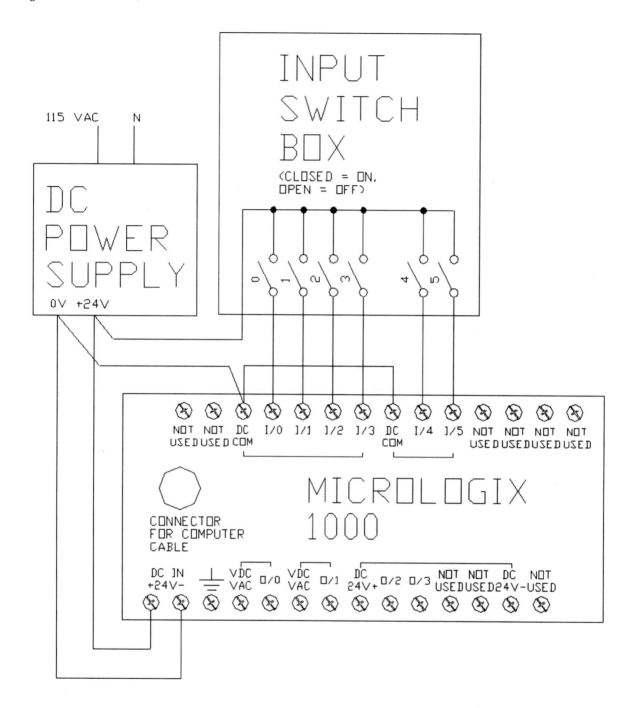

Figure 3-2 Schematic diagram of complete setup.

Bill of materials:

1) Wire, single conductor, 16 to 18 AWG
2) 115 VAC electrical wire and plug
3) Pine board,17.5" x 7.25" x .75"
4) Mounting screws
5) Wire ties
6) Power Supply (see below for details)
7) Output Switch Box (see below for details)

Construction note:
   Space should be left on the pine wood board to mount output devices, if later desired.

3.1 MICROLOGIX 1000

The MicroLogix 1000 Model 1761-L10BXB 10-point controller was used. It is necessary to use a 10-point controller rather than one with more points (input and output connection points) to make use of the free RSLogix 500 software. I purchased the 1761-L10BXB new from an electrical supply company for about $80.00. It is the least expensive Allen-Bradley programmable controller.

The 1761-L10BWA and 1761-L10BWB are also 10-point controllers that work with the free RSLogix 500 software. The 1761-L10BWB and 1761-L10BWA have only relay outputs. This is different from the 1761-L10BXB that has two MOSFET outputs and two relay outputs. The 1761-L10BWA has the advantage of not requiring a low voltage DC power supply. It is powered directly by 85 to 264 volt AC at 47 to 63 Hz.

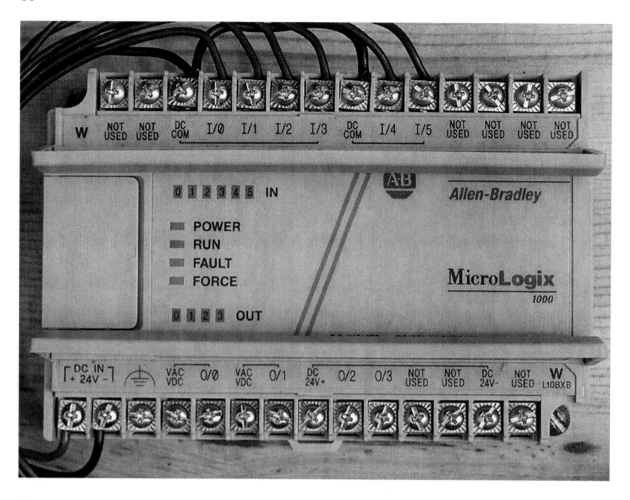

Figure 3-3 Photo that shows MicroLogix 1000 connection points.

3.2 MICROLOGIX 1000 TO PERSONAL COMPUTER CABLE

The MicroLogix 1000 is connected to a personal computer with a serial RS-232C cable. The Allen-Bradley part number of the appropriate cable is 1761-CBL-PM02.

To save money, I purchased the cable on eBay for about $20.00, including shipping.

The cable I purchased and the Allen-Bradley cable have 9 pin D-Sub female connectors on the computer end of the cables. Some older computers need a 9 pin D-Sub male to 25 pin D-Sub port adapter to connect. Newer computers, that have only USB inputs, will require a USB to RS232 serial pin cable adapter. These adapters can be bought on eBay and at other places.

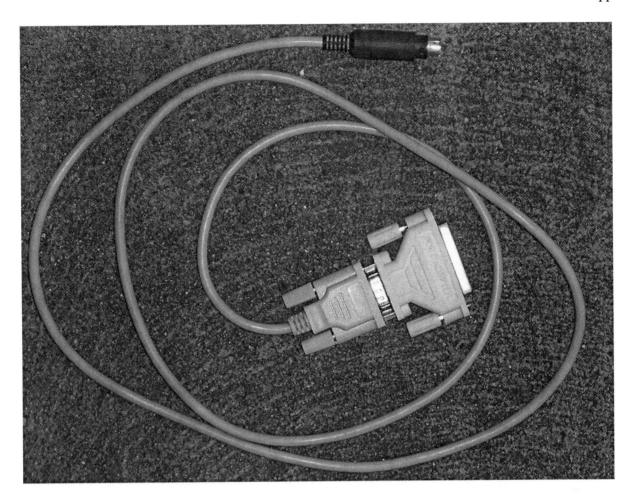

Figure 3-4 Photo of MicroLogix 1000 to personal computer connection cable with a 9 pin D-Sub male to 25 pin D-Sub port adapter.

3.3 POWER SUPPLY

The MicroLogix 1000 1761-L10BXB requires a DC input voltage of 20.4 to 26.4 volts.

I built a regulated DC supply to do this.

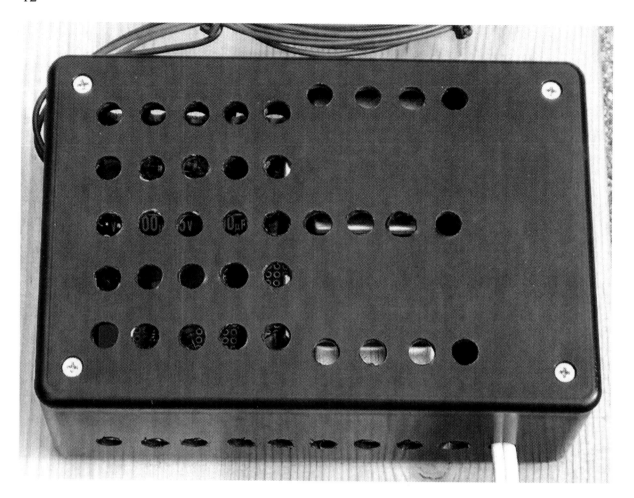

Figure 3-5 Photo of closed 24 VDC regulated power supply.

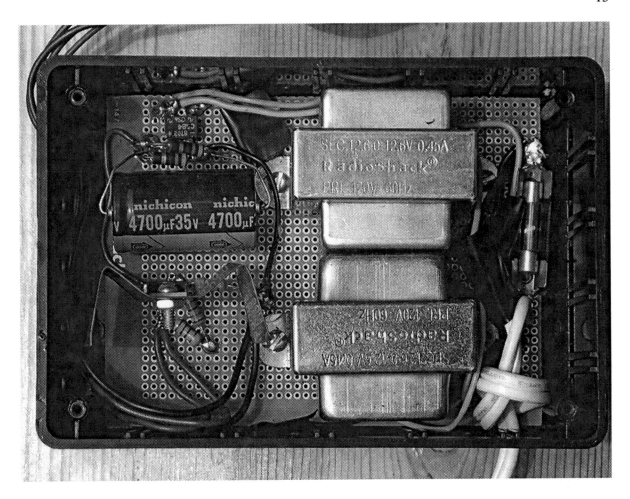

Figure 3-6 Photo of open 24 VDC regulated power supply.

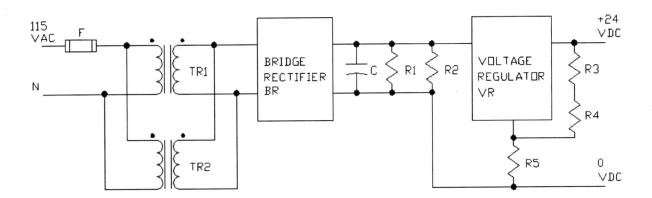

Figure 3-7 Schematic diagram of the 24 VDC regulated power supply.

Bill of materials:

1) Plastic mounting box, 4" x 2.125" x 6".
2) Perforation circuit board, 3.75" x 5".
3) Fuse holder for 1" glass tube fuse.
4) Fuse, 1" long glass tube, 250V, .5 A. (F)
5) Two 120/25.2V .45A output transformers. (TR1 & TR2)
6) Electrolytic capacitor, 4700 microF, 35 VDC. (C)
7) Bridge rectifier, 8702 CSB4, (Max. 400 PIV and max. 1 A average output on this device, a lower PIV higher current device could be used) (BR)
8) Two resistors, 4700 ohms +/- 10%. (R1 & R2)
9) Resistor, 4700 ohms +/- 5%. (R3)
10) Resistor, 150 ohms +/- 5%. (R4)
11) Resistor, 100 ohms +/- 5%. (R5)
12) Voltage regulator, LM317T PM19AN, (output voltage 1.3 to 37.5 V, maximum output current 1.5 A) (VR)
13) Copper strip heat sink.

Construction notes:

a) Solder all parts onto and through the perforated board.
b) Holes were drilled in the plastic box to allow cooling air to go past the voltage regulator chip and transformer.
c) All parts were purchased from Radio Shack.

3.4 INPUT SWITCH BOX

The MicroLogix 1000 1761-L10BXB has 6 digital inputs. These inputs are connected to a 20.4 to 26.4 DC positive voltage through two-position switches.

I built a switch box with 6 slide switches to do this.

Figure 3-8 Photo of closed input switch box.

Figure 3-9 Photo of open input switch box.

The schematic diagram of the input switch box is included in Figure 3-2.

Bill of materials:
    1) Plastic box, 1.125" x 4" x 2".
    2) White out to paint on switch numbers.
    3) Six 2-position switches, (SW0, SW1, SW2, SW3, SW4, SW5)

Construction note:
    All electrical parts may be purchased from Radio Shack.

4.0 RELAY LADDER AND POWER DIAGRAMS

Relay ladder diagrams show electrical relays, contactor coils, and indicator lights in control circuits. Power diagrams show the electrical power wiring to motors and other power equipment.

Relay ladder diagrams look like ladders. They have conductors on the left and right like the rails of a ladder and they have circuits connecting those rails that look like rungs.

Figure 4-1 is a one rung ladder diagram. The pushbutton and relay contacts "A" are the control elements. The relay coil "B" is controlled. The relay contacts "A" will close when relay "A" receives voltage. The coil for relay contacts "A" is not shown in Figure 4-1. When the pushbutton is pressed and the relay contacts "A" are closed a conducting path is made to the relay coil "B". The connection of the relay coil "B" to voltage causes it to operate. Relay contacts for coil "B" are not shown in Figure 4-1.

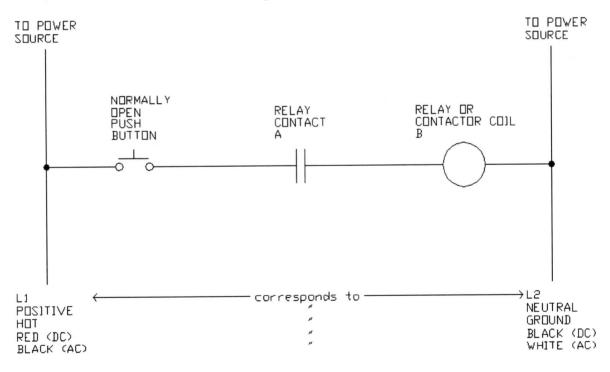

Figure 4-1 One rung ladder diagram.

More complicated ladder diagrams use more circuit symbols. Appendix 7.1, Relay Ladder and Power Diagram Symbols, shows commonly used symbols.

Following is an example of a control circuit that will run a three-phase induction motor in forward and reverse. The circuit is shown in a ladder diagram circuit in Figure 4-2 and a power diagram in Figure 4-3. Notice that if the circuit breaker trips open or the power disconnect is opened the ladder diagram circuit will be de-energized and thereby reset to the "stop" condition. On many control circuits the FORWARD contactor and REVERSE contactor are mechanically interlocked to make it impossible for a malfunction to close both motor contactors at one time.

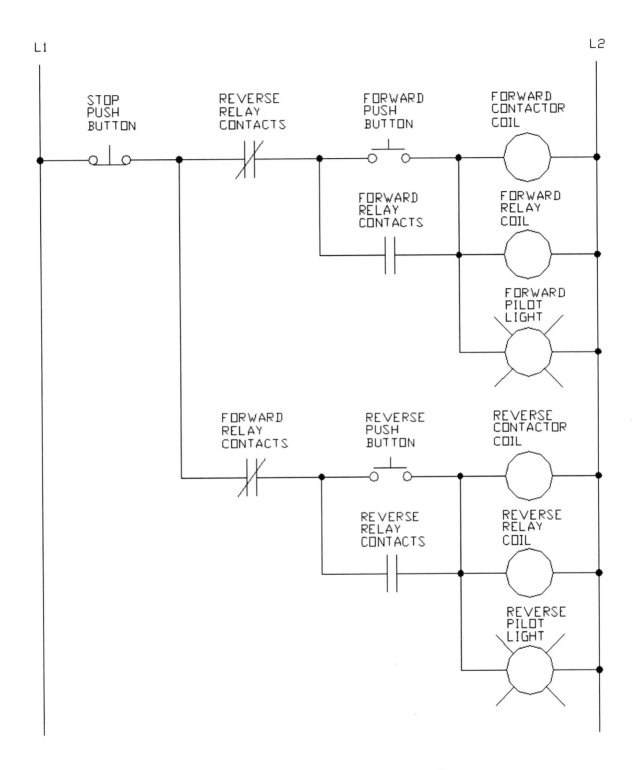

Figure 4-2 Forward and reverse motor control relay logic ladder diagram.

20

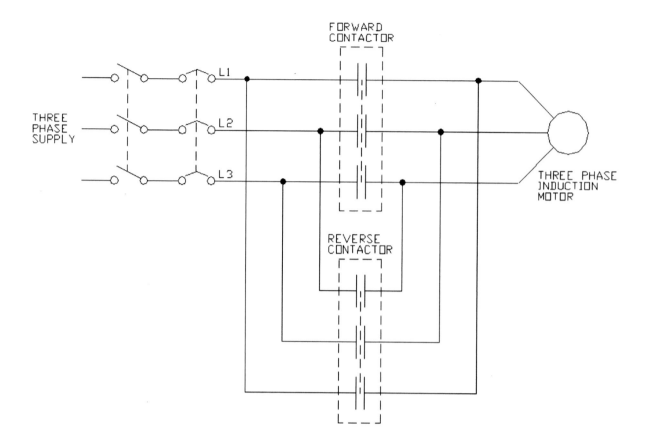

Figure 4-3 Forward and reverse motor control power diagram.

5.0 EXAMPLE MICROLOGIX 1000 LADDER AND POWER DIAGRAMS

The examples are designed to demonstrate the use of the more important MicroLogix 1000 and RSLogix 500 ladder program functions and details. For clarity, the example ladder diagrams are short, using less than one page each. Actual industrial ladder diagrams would probably be longer.

Each example contains a ladder diagram program and MicroLogix 1000 circuit connection diagram. Most also contain a power circuit connection diagram. The student can program and operate the ladder diagram programs with only the "MICROLOGIX 1000 TEACHING SETUP". It is not necessary to build the MicroLogix circuits and power circuits. Those circuits are only presented to clarify what the MicroLogix 1000 is designed to do in each example.

5.1 MICROLOGIX 1000 LADDER DIAGRAM AND POWER DIAGRAMS FOR THREE-PHASE INDUCTION MOTOR FORWARD AND REVERSE CONTROL

The relay logic scheme of chapter 4.0 is redone using the simplest possible MicroLogix 1000 ladder diagram and associated schematic diagrams.

The new material demonstrated here is the use of:
1) RSLogix 500 and RSLinx.
2) "Examine Open" and "Examine Closed" ladder diagram contacts.
3) Ladder diagram outputs.
4) MicroLogix 1000 wiring diagrams.

5.1.1 Using RSLogix 500 to load the ladder diagram program into a personal computer.

5.1.1.1 Start RSLogix 500 Starter for 10-pt MicroLogix 1000.

5.1.1.2 Go to File and click on New.

5.1.1.3 A window titled "Select Processor Type" comes up.

22

5.1.1.4 Already made selections that appear in the window are:
    5.1.1.4.1 Processor Name "UNTITLED"
    5.1.1.4.2 "Bul. 1761 MicroLogix 1000 DH-485/DHSlave"
    5.1.1.4.3 Driver "AB_DF1-1"
    5.1.1.4.4 Processor Node: "1"
    5.1.1.4.5 Reply Timeout: "10"

5.1.1.5 Change "UNTITLED" to "Figure 5-5".

5.1.1.6 Click "OK".

5.1.1.7 Two windows will come up. One is titled "Figure 5-5". The other is titled "LAD 2 – MAIN_PROG". The ladder diagram page is shown in Figure 5-1.

Figure 5-1 Blank ladder diagram window.

5.1.1.8 To draw the ladder diagram.
    5.1.1.8.1 Click the "New Rung" icon above the ladder diagram window.
    5.1.1.8.2 Pull down a new rung and place it above rung 0000. The icon should be brought close to the 0000 and a small green square will appear on the screen. When the green square appears the left mouse button can be released and the rung will go to the proper position. The green squares are attachment points. They appear when a function symbol is dragged near them.
    5.1.1.8.3 Pull down an "Examine if Closed" symbol and place it on the top rung just to the right of a green square on the left of the 0000 rung.

Figure 5-2 Ladder diagram window with an "Examine if Closed" symbol added.

5.1.1.8.4 The "Examine if Open", "Examine if Closed", "Output Energize", and "Rung Branch" symbols can be pulled down to create the ladder diagram of Figure 5-3. It follows the general design of the relay ladder diagram of Figure 4-3.

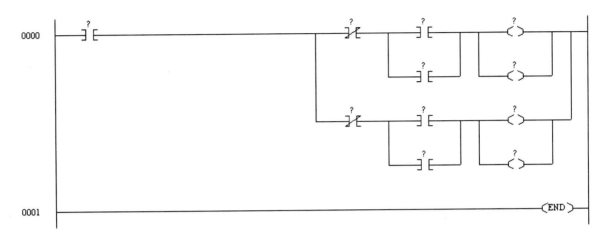

Figure 5-3 Ladder diagram window with all needed functions.

5.1.1.8.5 Add the address I:0/0 to the upper left "Examine if Open" symbol and give it the label "STOP". This is done by first clicking on the "Examine if Open" symbol to select it. Then double click on the question mark. A small window will appear where I:0/0 and <CR>should be typed in. Then another window will appear that asks for a description and symbol. Type "STOP" in as the description and leave the symbol box empty. This is shown in Figure 5-4.

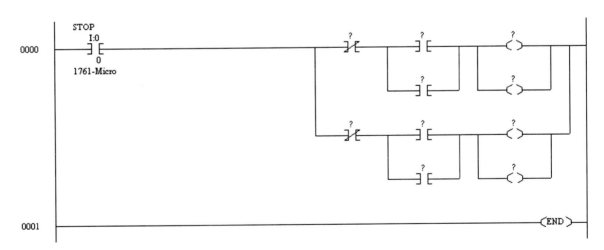

Figure 5-4 Ladder diagram with one addressed and described function symbol.

5.1.1.8.6 Type the rest of the addresses and descriptions as shown in Figure 5-5. This completes the creation of the ladder diagram.

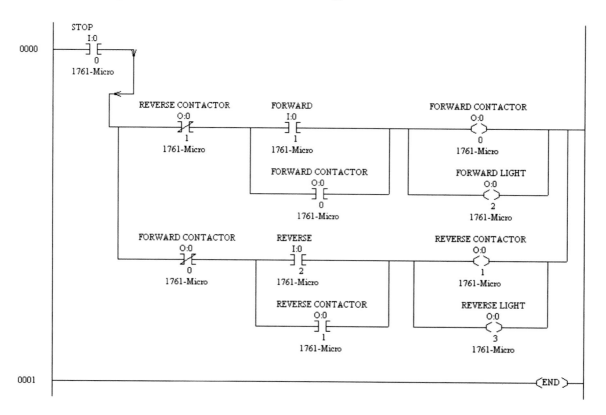

Figure 5-5 Ladder diagram with addressed and described function symbols.

5.1.2 Validating the ladder diagram program.

Before compiling and downloading your ladder diagram to the MicroLogix 1000 it should be validated. This is done by clicking Edit>Verify file. The Verify Results output page will give information about mistakes or omissions that may be in your ladder diagram.

5.1.3 MicroLogix 1000 hard-wired circuit.

The MicroLogix 1000 hard-wired circuit can be made with or without compliance with National Electrical Manufacturing Association (NEMA) recommendations. NEMA recommends, *"When the operator is exposed to the machinery, such as loading or unloading a machine tool, or where the machine cycles automatically, consideration should be given to the use of an electromechanical override or other redundant means, independent of the controller, for starting or interrupting cycle"*. Following the NEMA recommendation means that the STOP pushbutton in our motor forward and reverse controller should be hardwired into the motor control circuit so that the motor will stop when the STOP pushbutton is pushed regardless of whether the MicroLogix 1000 functions or not. Two power diagrams will be presented one with and one without compliance.

In both circuits MicroLogix 1000 power is received after the main circuit breaker and disconnect. This will cause the MicroLogix 1000 to reset to the "stop" condition soon after the main circuit breaker or disconnect opens.

5.1.3.1 MicroLogix 1000 circuit that doesn't comply with the NEMA recommendation.

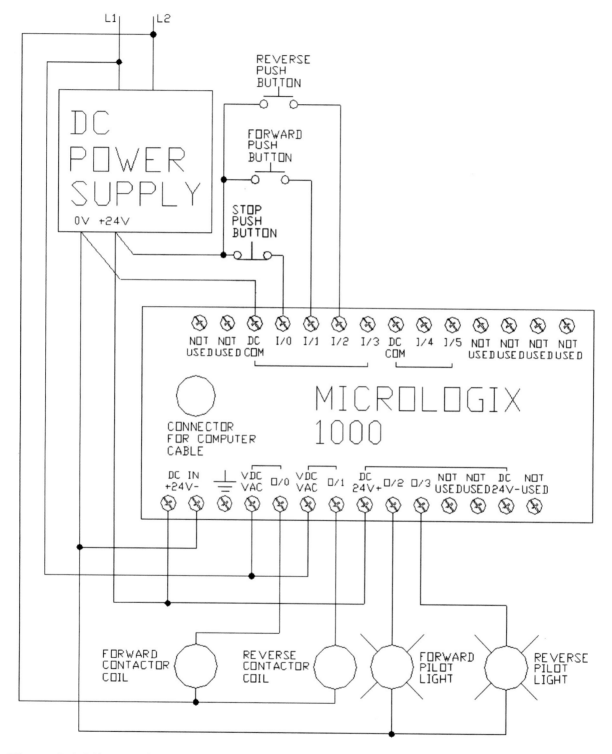

Figure 5-6 MicroLogix 1000 circuit without NEMA compliance.

5.1.3.2 MicroLogix 1000 circuit that complies with the NEMA recommendation.

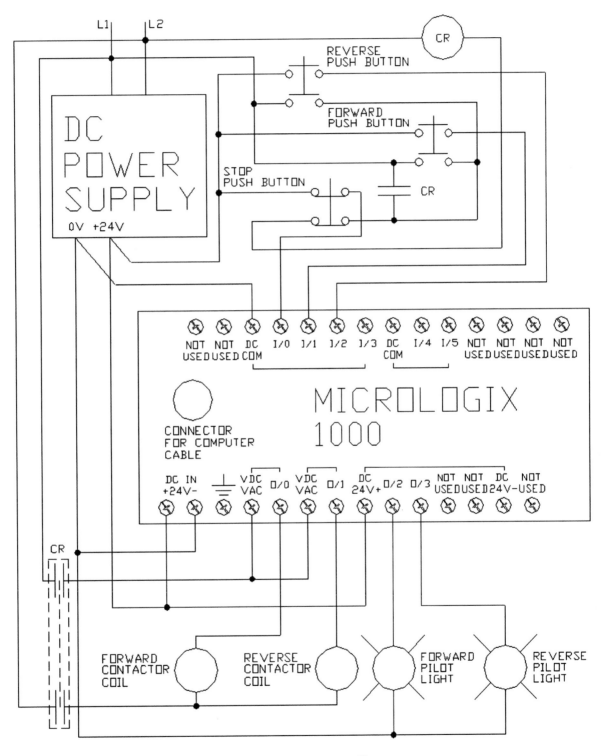

Figure 5-7 MicroLogix 1000 circuit with NEMA compliance.

5.1.4 Power circuit.

The same power circuit is used here as was used in Chapter 4.0 with the relay ladder diagram. See Figure 4-3.

5.1.5 Preparing the personal computer for downloading the ladder diagram program from RSLogix 500 to the MicroLogix 1000.

5.1.5.1 Connect the cable from the personal computer to the MicroLogix 1000.

5.1.5.2 Apply power to the MicroLogix 1000.

5.1.5.3 Start the RSLinx Lite program.

5.1.5.4 A page appears that is titled "RSLinx Lite".

5.1.5.5 Configure a driver for RSLinx Lite and your computer. This should only have to be done once.
5.1.5.5.1 Go to "Communications" and click "Configure Drivers".
5.1.5.5.2 Click on "Available Driver Types" and select "RS-232 DF1devices".
5.1.5.5.3 Click "Add New".
5.1.5.5.4 Click "OK" on "AB_DF1-1".
5.1.5.5.5 Click "Autoconfigure" in the "Configure RS-232DF1 Devices" and then click "OK".
5.1.5.5.6 A window should come up that says, "Name and Description = AB_DF1-1 DH485 Sta: 0COM1: RUNNING, Status = Running".
5.1.5.5.7 Click "Close".

5.1.5.6 Click on "Communications" and then click on "RSWho".

5.1.5.7 Click on the icon titled "AB_DF-1 DH-485".
5.1.5.7.1 A new window will come up, although it still has the same title "RSLinx Lite – [RSWho – 1]".
5.1.5.7.2 Click on the icon that looks like a computer and is titled "00 DF1-COM1".
5.1.5.7.3 Leave the RSLinx Lite program running and push its window to one side of the screen.

5.1.6 Downloading the ladder diagram program.

5.1.6.1 Start RSLogix 500 again and open the ladder diagram "Figure 5-5"

5.1.6.2 Go to the "Comms" heading in RSLogix 500.

5.1.6.3 Under the "Comms" heading click on the "Download" option.

5.1.6.4 Follow the prompts. The ladder diagram should download via RSLinx Lite to the MicroLogix 1000. and run. Choose to run in the "On-line mode" to see connections being made on the screen as they are happening in the MicroLogix 1000.

5.1.7 Running the ladder program with the teaching setup.

The operation of the MicroLogix 1000 teaching setup will verify proper operation of the RSLogix 500 ladder diagram.

The MicroLogix 1000 has indicating LEDs on both its input and output. The input status can also be determined by the switch position. The output status can be checked with the output LEDs. It isn't necessary to build the circuits of Figures 5-6, 5-7, and 4-3.

The operation of the ladder diagram can also be observed through the combined use of RSLogix 500 and RSLinx while the MicroLogix 1000 is connected by its interface cable to your personal computer. When the ladder diagram was being downloaded to the MicroLogix 1000, prompts asked if you wanted the MicroLogix 1000 in the "Run" mode and in the "Online" mode. If you answered yes to these you will see a visual indication of how switching the MicroLogix 1000 inputs causes the ladder diagram to operate. When in the online run mode contacts can be seen closed by green rectangles on each side of the closed contacts. Running information can also be seen in the various Data Files. For example, the I1 Input Data File can be clicked on and the different inputs watched as the MicroLogix 1000 is in the "Online" and "Run" mode. Figure 5-8 below shows the ladder program of Figure 5-5 in the run mode with switch 0 and switch 2 closed and switch 1 open.

30

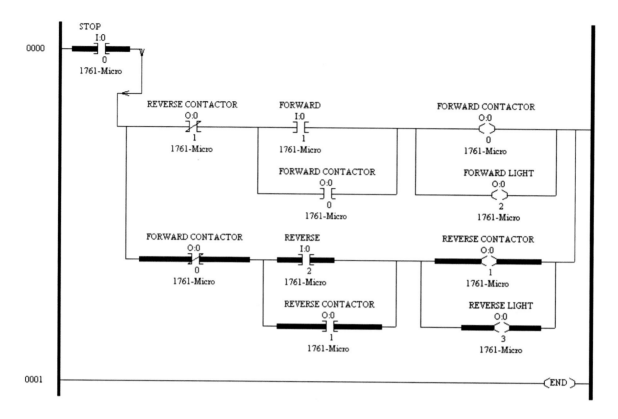

Figure 5-8 RSLogix 500 ladder diagram program running in the online mode.

5.1.6 Worth of using a MicroLogix 1000 to perform the forward and reverse switching.

Is it sensible to use a MicroLogix 1000 to perform the forward and reverse switching of a motor rather than the simple relay ladder circuit of Chapter 4.0? No, it isn't. Counting and comparing parts one sees that the MicroLogix 1000 only replaces two relays. Probably those two relays would be less expensive than the MicroLogix 1000 and the relays don't need programming software. The reason this MicroLogix 1000 circuit was put here was instructive, rather than practical. More complicated relay circuits involving many more relays and counters and timers would better justify the use of the MicroLogix 1000.

5.2 MICROLOGIX 1000 LADDER DIAGRAM AND POWER DIAGRAMS FOR TIMED SEQUENTIAL STARTING OF TWO THREE-PHASE INDUCTION MOTORS

In this example the cutting oil pump on a milling machine is controlled so that it starts and runs before and after the machine cuts metal. This assures that no cutting is done with an un-lubricated and un-cooled cutting tool. The MicroLogix 1000 does this by controlling power to a cutting oil pump motor and main milling machine drive motor. This is done without the NEMA recommended hardwired STOP pushbutton to allow us to focus our learning on the MicroLogix 1000 operation.

The new material demonstrated here is the use of:
1) The "Time On Delay" function.
2) The "Output Latch" function.
3) The "Output Unlatch" function.

Operation sequence with the ladder diagram of Figure 5-9:
1) All systems off.
2) Circuit breaker disconnect switch closed. Power goes to the MicroLogix 1000.
3) The START pushbutton is pressed. This unlatches the STOP, starts the oil pump motor, and starts the "T4:0 Timer On Delay".
4) After a 3 second delay the "T4:0 Timer On Delay" starts the main drive motor.
5) Now the main drive motor and oil pump motor run simultaneously.
6) The STOP pushbutton is pressed latching STOP on.
7) The main drive motor has power removed immediately and the "T4:1 Timer On Delay" starts.
8) After a 3 second delay the "T4:1 Timer On Delay" removes power from the oil pump motor.

32

5.2.1 Ladder diagram.

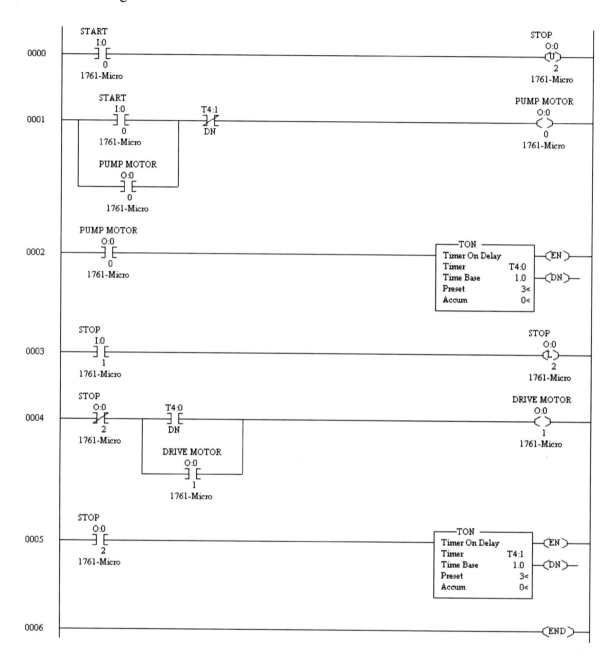

Figure 5-9 Ladder diagram for the milling machine cutting oil motor/drive motor controller.

5.2.2 MicroLogix 1000 circuit.

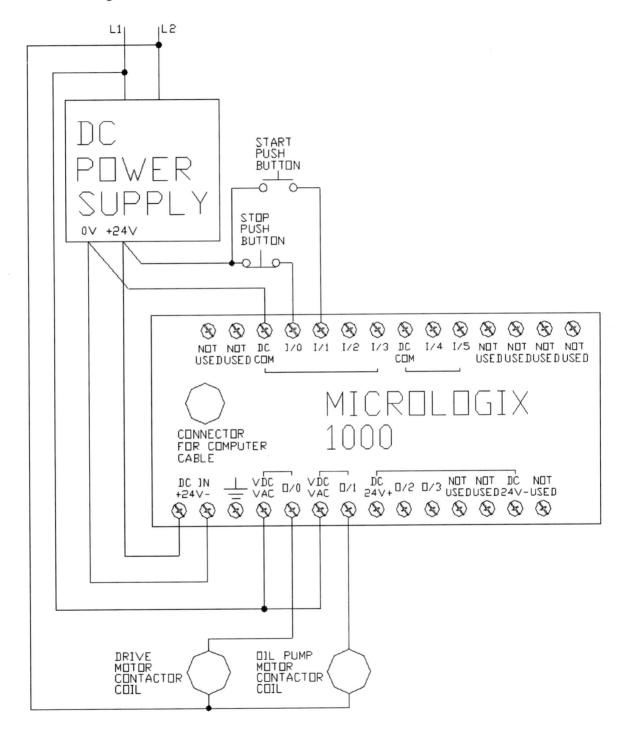

Figure 5-10 MicroLogix 1000 circuit for the milling machine cutting oil motor/drive motor controller.

34

5.2.3 Power circuit.

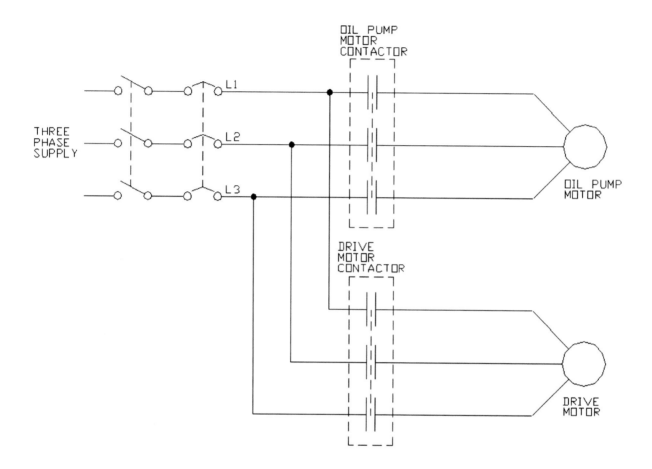

Figure 5-11 Oil pump motor and drive motor power diagram.

5.3 MICROLOGIX 1000 LADDER DIAGRAM AND POWER DIAGRAMS FOR A COUNTER CONTROLLED THREE-PHASE INDUCTION MOTOR POWERED BOTTLE PUSHER

In this example bottles on a packaging machine conveyor belt are pushed off to one side after six have passed. The pusher is made to move by a three-phase induction motor. After the pusher has gone to its push limit the motor reverses and the pusher returns to await six more bottles. This operation might be done in a bottling plant where the bottles are being dropped into boxes.

The new material demonstrated here is the use of:
1) The "Count Up" function.
2) The "Reset" function.

Operation sequence with the ladder diagram of Figure 5-12:
1) All systems off.
2) Circuit breaker disconnect switch closed. Power goes to the MicroLogix 1000.
3) The pusher motor is at rest in the retracted position.
4) One bottle goes into the push area.
5) The counter limit switch, BOTTLE INDICATOR, opens and closes once as the bottle passes.
6) The opening and closing of BOTTLE INDICATOR causes the Count Up BOTTLE COUNTER to increase the value in C5:0 by 1.
7) Five more bottles go into the push area. The counter limit switch opens and closes five more times. This causes C5:0 to increase by 5 more to a total of 6.
8) Since C5:0 has reached its preset value of 6 the BOTTLE COUNTER is full and will now close the logical contact C5:0/DN.
9) The closing of logical contact C5:0/DN causes the FORWARD TRAVEL output to go on, pushing all 6 bottles off the conveyor belt.
10) After the motor has pushed to its limit, so as to open the FORWARD LIMIT SWITCH, the motor reverses and draws back the pusher.
11) When the pusher has returned to its retracted position the REAR TRAVEL limit switch opens, the motor shuts off, and the counter is reset.
12) The motor and counter are now ready to begin again.

36

5.3.1 Ladder diagram for the bottle pusher controller.

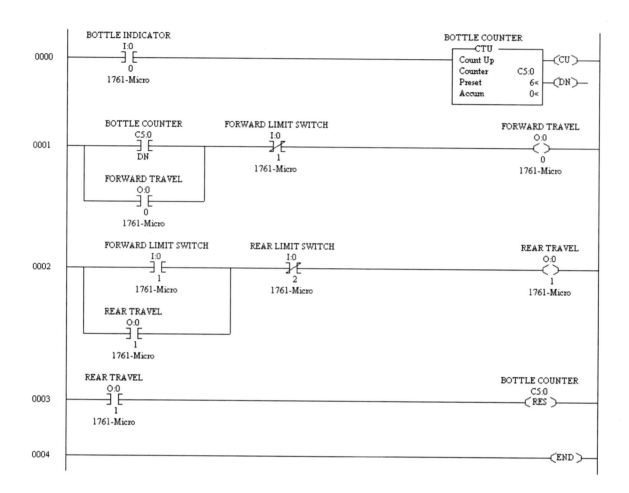

Figure 5-12 Ladder diagram for the bottle pusher controller.

5.3.2 MicroLogix 1000 circuit for the bottle pusher.

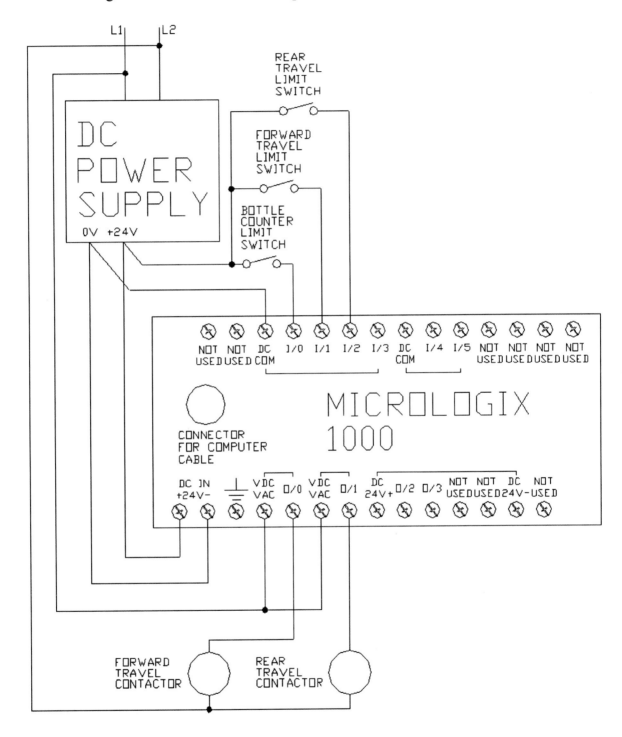

Figure 5-13 MicroLogix 1000 circuit for the bottle pusher controller.

5.3.3 Power circuit.

The same power circuit is used here as was used in Chapter 4.0 with the relay ladder diagram. See Figure 4-3.

5.4 MICROLOGIX 1000 LADDER DIAGRAM AND POWER DIAGRAMS FOR A COUNTER CONTROLLED THREE-PHASE INDUCTION MOTOR WITH A SETABLE COUNTER

This example is the same as 5.3 except that a manual switch has been added that allows an operator to change the counter from six to eight bottles.

The new material demonstrated here is the use of the "Move" function.

Operation sequence with the ladder diagram of Figure 5-14:

1) All systems off.

2) Circuit breaker disconnect switch closed. Power goes to the MicroLogix 1000.

3) If the 6 BOTTLES ON 8 BOTTLES OFF switch, I:0/3, is closed, operation is identical to that in section 5.3.

4) If the 6 BOTTLES ON 8 BOTTLES OFF switch, I:0/3, is open the BOTTLE COUNTER Move will place 8 into the Preset of the BOTTLE COUNTER Count Up.

5) The program will now count out 8 bottles rather than 6, but will otherwise be the same as that in section 5.3.

40

5.4.1 Ladder diagram for the bottle pusher controller with a settable counter.

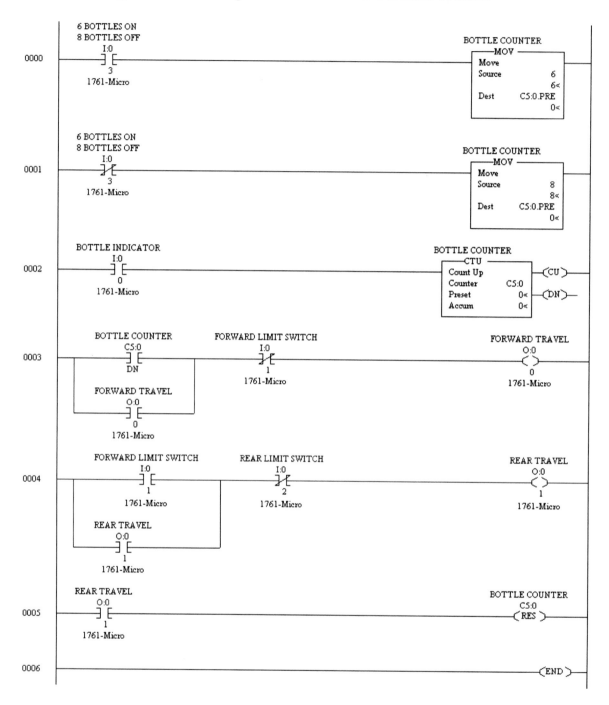

Figure 5-14 Ladder diagram for the bottle pusher controller with settable counter.

5.4.2 MicroLogix 1000 circuit for the bottle pusher with a settable counter.

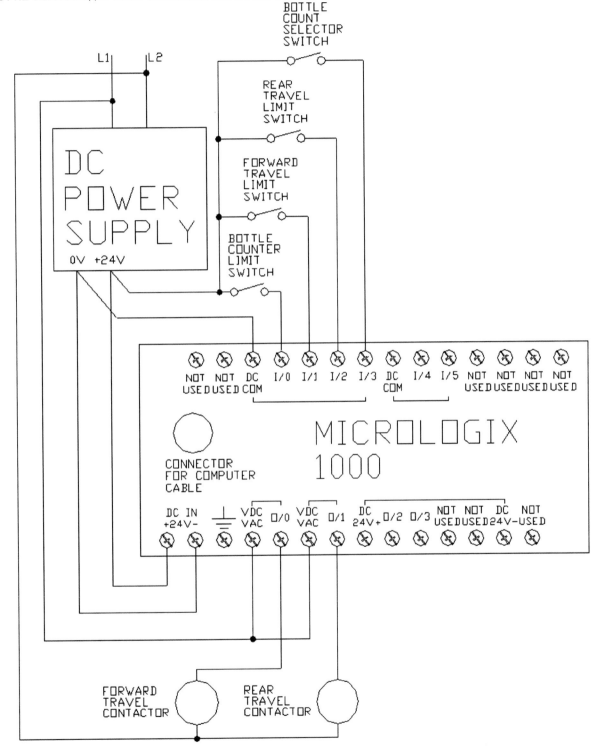

Figure 5-15 MicroLogix 1000 circuit for the bottle pusher controller with settable counter.

5.4.3 Power circuit.

The same power circuit is used here as was used in Chapter 4.0 with the relay ladder diagram. See Figure 4-3.

5.5 MICROLOGIX 1000 LADDER DIAGRAM AND POWER DIAGRAMS FOR ANOTHER COUNTER CONTROLLED THREE-PHASE INDUCTION MOTOR

In this example bottles coming from a conveyor belt are accumulated in a holding area. If less than six bottles are in the area, the three-phase induction motor powered input conveyor belt is started. Once six bottles are accumulated, the power to the induction motor powered conveyor is turned off.

There is a limit switch that closes momentarily every time a bottle goes into the accumulator area and another that closes momentarily every time a bottle goes out of the accumulator area.

The new material demonstrated here is the use of the "Count Down" function.

Operation sequence with the ladder diagram of Figure 5-16:
1) There are no bottles in the accumulator area.
2) The input conveyor belt brings in six bottles. Each bottle is counted and the total number stored in C5:0.
3) The conveyor belt stops when C5:0 has counted up to 6 since the accumulator is now full.
4) The accumulator sends out bottles. As each bottle goes out 1 is subtracted from the total in C5:0. When there are less than six bottles in the accumulator area, the conveyor belt starts up.
5) Once the accumulator has six bottles in it again the input conveyor belt stops again.

44

5.5.1 Ladder diagram for the bottle accumulator.

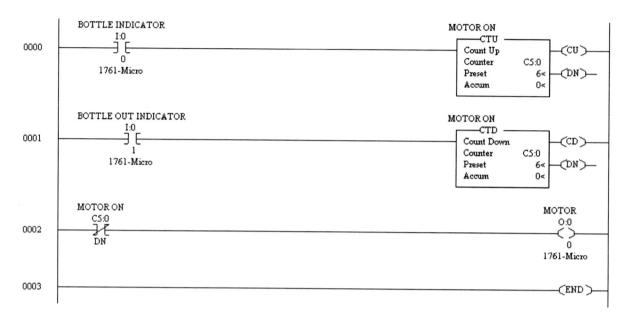

Figure 5-16 Ladder diagram for the bottle accumulator.

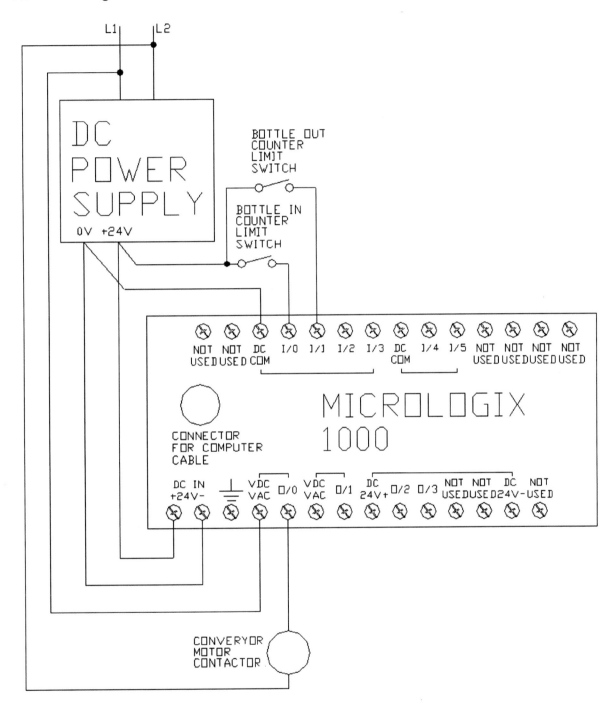

5.5.2 MicroLogix 1000 circuit for the bottle accumulator.

Figure 5-17 MicroLogix 1000 circuit for the bottle accumulator.

46

5.5.3 Power circuit for the bottle accumulator.

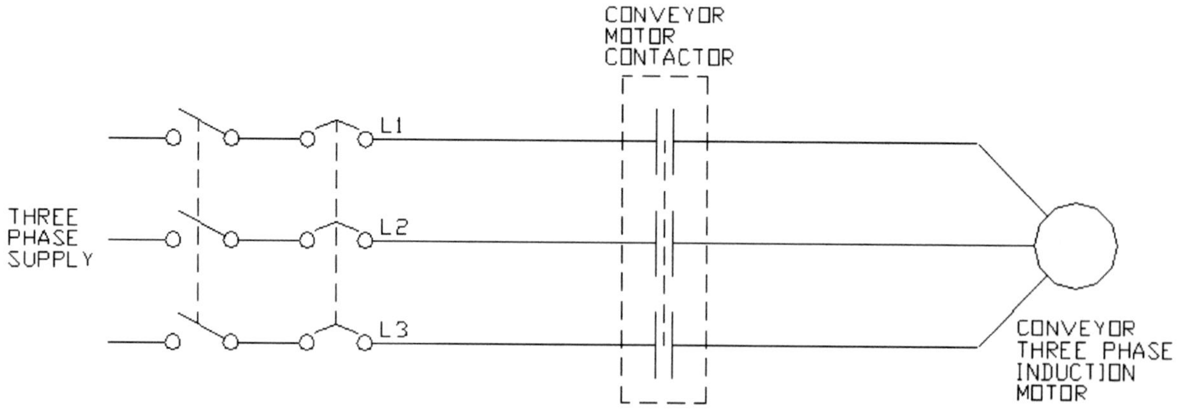

Figure 5-18 Bottle accumulator conveyor motor power diagram.

5.6 MICROLOGIX 1000 LADDER DIAGRAM AND POWER DIAGRAMS FOR TIMED SEQUENTIAL STARTING OF TWO THREE-PHASE INDUCTION MOTORS WITH A MASTER CONTROL RESET INSTANT STOP

This is a modification of the example of section 5.2. In section 5.2 the oil pump motor can not be immediately shutdown, there is a 3 second time delay after the STOP pushbutton is pressed. This section's circuit includes a master control reset switch that instantly removes power to the milling machine cutting motor and the cutting oil pump motor. As in section 5.2 this will be done without the NEMA recommended hardwired STOP pushbutton.

The new material demonstrated here is the use of the "Master Control Reset" function. The "Master Control Reset" function turns off all outputs in a ladder diagram between the MCR outputs when the upper MCR output is not enabled.

Operation sequence of the ladder diagram of Figure 5-19, with MCR INSTANT STOP not being pressed (pushbutton contacts closed), is the same as Figure 5-9. See section 5.2.

Operation sequence of the ladder diagram of Figure 5-19, when MCR INSTANT STOP being is pressed:
1) All systems off.
2) Circuit breaker disconnect switch closed. Power goes to the MicroLogix 1000.
3) The START pushbutton is pressed.
4) The oil pump motor receives power and starts.
5) After a 3 second delay the main drive motor receives power and starts.
6) The main drive motor and oil pump motor run simultaneously.
7) The MCR INSTANT STOP pushbutton is pressed.
8) The outputs controlling the main drive motor and oil pump motor will have power removed immediately and the timers will reset to time 0.
9) Letting the MCR INSTANT STOP pushbutton go back to its normally closed position makes the ladder diagram program available for normal operation again.

48

5.6.1 Ladder diagram.

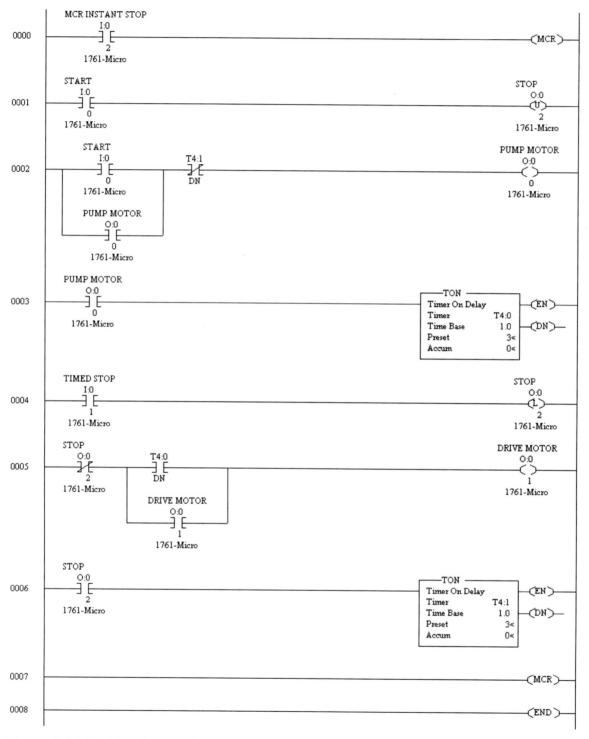

Figure 5-19 Ladder diagram for the milling machine cutting oil motor/drive motor controller with an instant stop.

5.6.2 MicroLogix 1000 circuit.

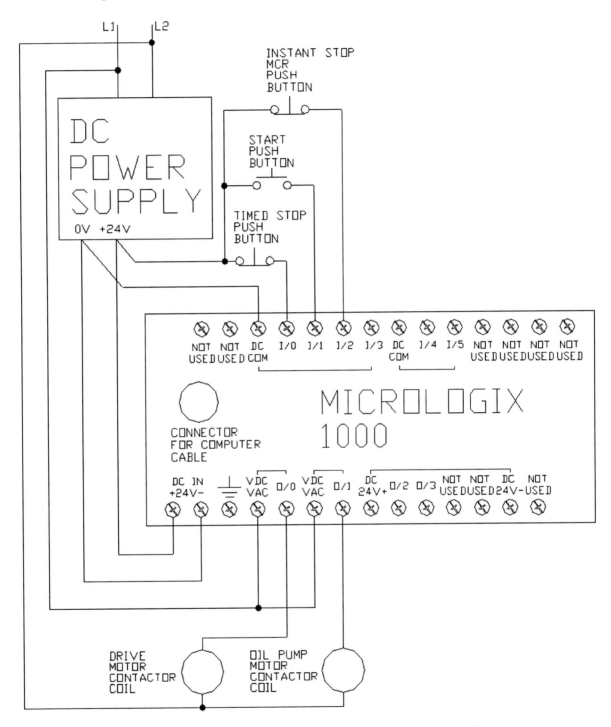

Figure 5-20 MicroLogix 1000 circuit for the milling machine cutting oil motor/drive motor controller with an instant stop.

5.6.3 Power circuit.

The same power circuit is used here as was used in Chapter 5.2. See Figure 5-11.

5.7 MICROLOGIX 1000 LADDER DIAGRAM AND POWER DIAGRAMS FOR A RETENTIVE TIMER CONTROLLED MACHINE

In this example the MicroLogix 1000 stores a machine's total operating time. After the machine has operated for a set period of time a warning pilot light goes on and the machine is automatically shutdown.

The new material demonstrated here is the use of:
1) The "Retentive Timer On-delay" function.
2) The "cascading" of a timer with a counter.

Operation sequence with the ladder diagram of Figure 5-21:
1) All systems off.
2) Power goes to the MicroLogix 1000.
3) The MACHINE RUN TIME Accum (accumulator) has the value it was when the program was last run.
3) The MACHINE RUN pushbutton is pressed, starting the machine.
4) When the machine starts the MACHINE RUN TIME "Retentive Timer On" starts. In this case it counts in 1 second increments to a maximum of 5 seconds. The "Retentive Timer On" time is "cascaded" through a "Count Up" that multiplies the time by 2. In the ladder diagram this makes a resultant 5 x 2 = 10 second timer.
5) The cascading greatly extends the timing range of the MicroLogix 1000. The "Retentive Timer On" is capable of counting to a maximum of 32,767 seconds, which equals 9.10 hours. With the use of the "Count Up" set to its maximum of 32,767 counts the maximum timing range is 32,767 x 32,767 seconds = 34 years.
6) When the "Count Up" has reached its preset value the machine is shut down and a MAINTENANCE NEEDED pilot light goes on.
7) A RESET MACHINE RUN TIME pushbutton can be pressed. This turns off the MAINTENANCE NEEDED pilot light and puts the MicroLogix 1000 in a ready to start condition with zero time stored.

52

5.3.1 Ladder diagram for the retentive timer controlled machine

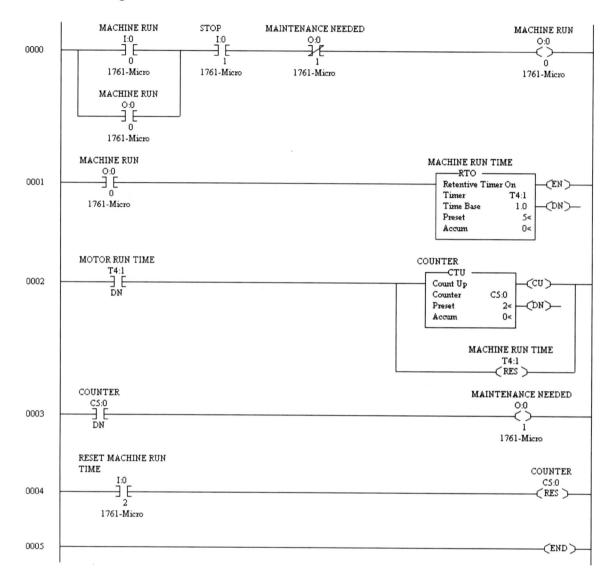

Figure 5-21 Ladder diagram for the retentive timer controlled machine.

5.3.2 MicroLogix 1000 circuit for the retentive timer controlled machine.

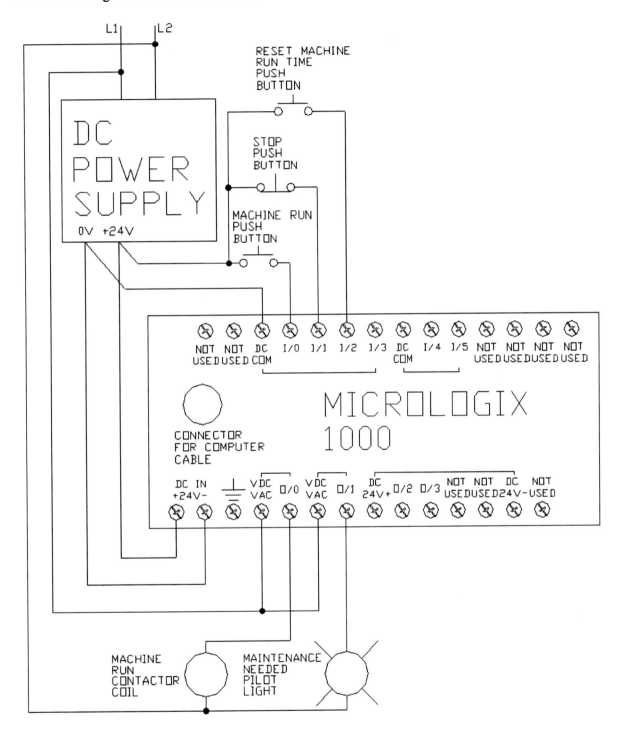

Figure 5-22 MicroLogix 1000 circuit for the retentive timer controlled machine.

5.8 MICROLOGIX 1000 LADDER DIAGRAM AND POWER DIAGRAMS FOR A SYSTEM TO DETERMINE IF A BOTTLE COUNT RATE IS HIGH OR LOW

In this example bottles riding on a conveyor belt are counted every 15 seconds. The number of bottles passing on the conveyor belt is compared to low and high level limits. If the number is less than 2 bottles per 15 seconds a low rate output pilot light is energized. If the number is greater than 4 bottles per 15 seconds a high rate output pilot light is energized.

The new material demonstrated here is the use of:
1) The "Equal to" function.
2) The "Less Than" function.
3) The "Greater Than" function.

Operation sequence with the ladder diagram of Figure 5-23:
1) All systems off.
2) Power goes to the MicroLogix 1000.
3) The BOTTLE COUNTER LIMIT SWITCH starts counting.
4) The COUNTING TIMER "Timer On Delay" and DISPLAY RESET TIMER "Timer On Delay" start timing in one second intervals.
5) After 14 seconds the DISPLAY RESET TIMER unlatches the low and high pilot light outputs, turning both off.
6) After 1 more second the COUNTING TIMER causes the number of counts accumulated by the BOTTLE COUNTER to be compared to the low limit of 2 and the high limit of 4. The comparisons are done by the BOTTLE COUNTER "Less Than" and "Greater Than" functions. If the counts are less than 2 the LOW LIMIT pilot light latches on. If the counts are greater than 4 the HIGH LIMIT pilot light latches on.
7) The "Equal to" COUNTING TIMER causes the counter and timers to be reset to zero and the cycle starting at 3) is ready to begin again.

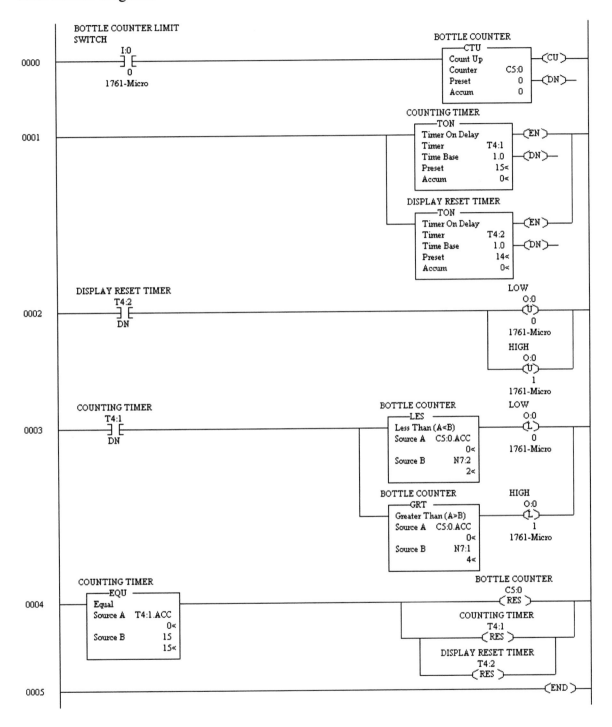

5.8.1 Ladder diagram.

Figure 5-23 Ladder diagram for the bottle rate checker.

56

5.8.2 MicroLogix 1000 circuit.

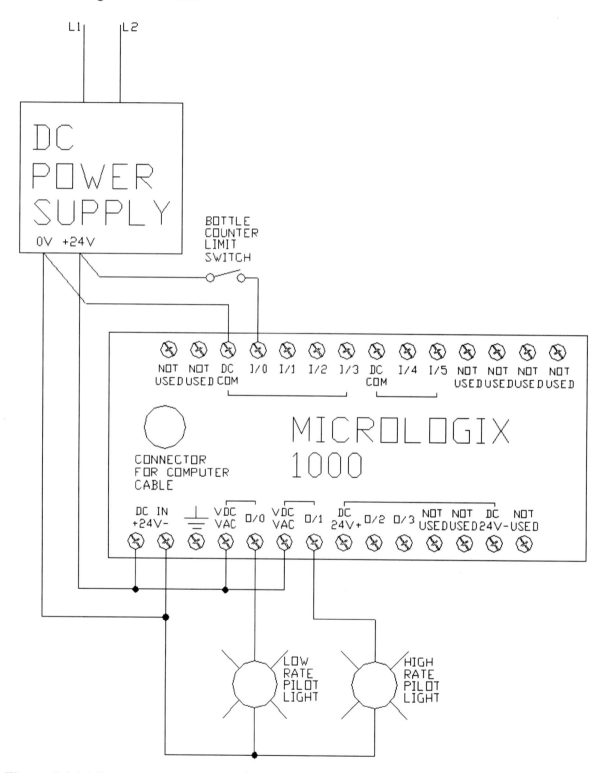

Figure 5-24 MicroLogix 1000 circuit for the bottle rate checker.

5.9 MICROLOGIX 1000 LADDER DIAGRAM AND POWER DIAGRAMS FOR A SYSTEM TO DETERMINE A BOTTLE COUNT RATE AND SELECT PRESET PUMP MOTOR INVERTER SPEEDS

In this example bottles are supplied by one conveyor belt at a constant rate of 3 bottles/15 seconds. Another conveyor belt supplies bottles at a variable rate between 0 and 4 bottles/15 seconds. The bottles of each conveyor belt are combined at the filling and capping machine. The combined bottle rate is used to select the pump motor speed for filling the bottles. The bottle rate selects one of two preset speeds on the pump motor inverter.

The new material demonstrated here is the use of:
1) The "Add" function.
2) The "Limit Test" function.

Operation sequence with the ladder diagram of Figure 5-25:
1) All systems off.
2) Power goes to the MicroLogix 1000.
3) The BOTTLE COUNTER LIMIT SWITCH starts counting.
4) The COUNTING TIMER starts timing in one second intervals.
5) After 15 seconds the counts accumulated by the BOTTLE COUNTER LIMIT SWITCH on the variable rate conveyor belt are added to those expected from the constant rate conveyor belt.
6) The total bottle count is compared to limits in the TOTAL BOTTLE COUNT "Limit Test" functions. The appropriate PRESET INVERTER SPEED is latched on and the other PRESET INVERTER SPEED latched off.
8) The BOTTLE COUNTER LIMIT SWITCH and COUNTING TIMER are both reset to zero.
9) The cycle of 3) repeats.

58

5.9.1 Ladder diagram.

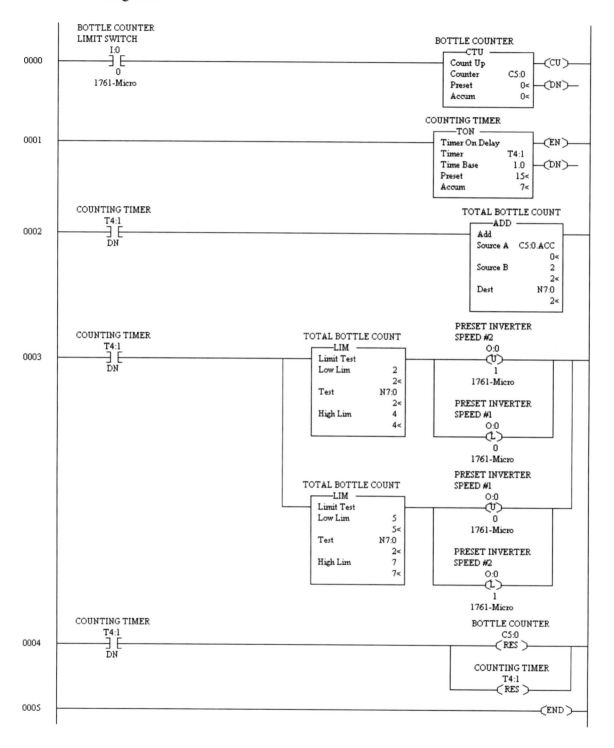

Figure 5-25 Ladder diagram for the bottle pump motor speed preset selector.

5.9.2 MicroLogix 1000 circuit.

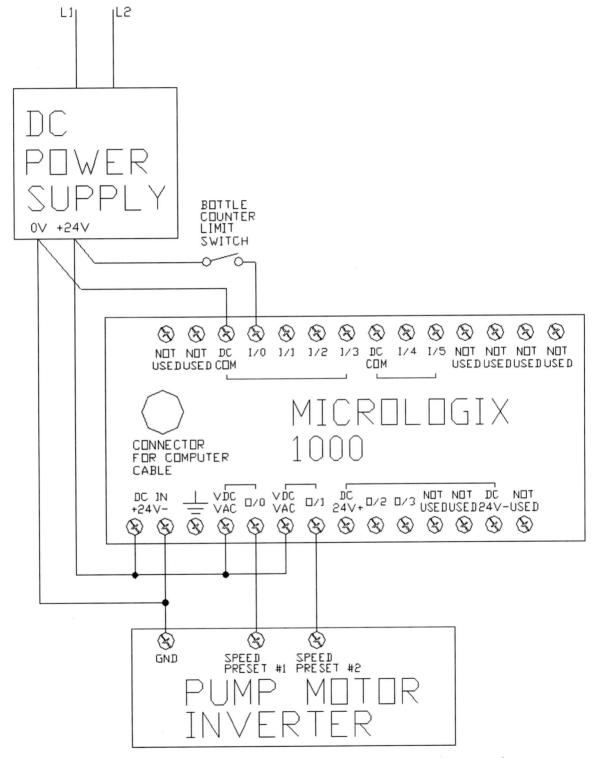

Figure 5-26 MicroLogix 1000 circuit for the bottle pump motor speed preset selector.

60

5.9.3 Power circuit.

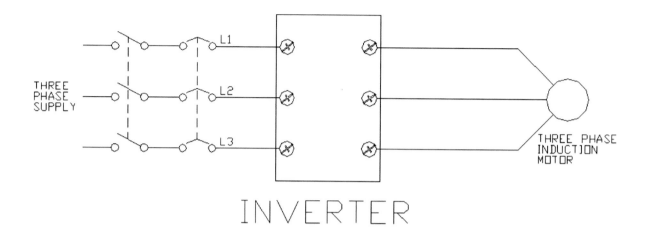

Figure 5-27 Bottle pump motor speed power diagram.

5.10 MICROLOGIX 1000 LADDER DIAGRAM AND POWER DIAGRAMS FOR A CONVEYOR BELT PART PLACER

In this example a conveyor belt is built with discrete trays. The trays may carry base parts or be empty. A machine places mating parts on the base parts. Five trays back from where the mating parts are added, there are two limit switches. One limit switch opens and closes each time a conveyor belt tray passes. The second limit switch opens and closes when a base part passes. The limit switch opening and closing information is remembered by the MicroLogix 1000. The MicroLogix 1000 uses the limit switch information of six tray locations back to decide whether or not it should place a mating part.

The new material demonstrated here is the use of the "Bit Shift Left" function.

Operation sequence with the ladder diagram of Figure 5-28:
1) The conveyor belt trays do not have base parts on them.
2) Power goes to the MicroLogix 1000.
3) The MicroLogix 1000 Bit Data register (word) ,B3:0, now looks like:

Offset	/5	/4	/3	/2	/1	/0
B3:0	0	0	0	0	0	0

4) The I:0/0, TRAY SENSING LIMIT SWITCH, closes and opens as each tray goes past.

5) The I:0/1, BASE PART SENSING LIMIT SWITCH, input places a 0 in the Bit Data Register at each count, because there are no base parts on the trays the I:0/1 limit switch does not operate.

6) The Bit Data register still looks like:

Offset	/5	/4	/3	/2	/1	/0
B3:0	0	0	0	0	0	0

7) The trays coming to the limit switch start to contain base parts so the BASE PART LIMIT SWITCH now closes and opens at the same time as the TRAY SENSING LIMIT SWITCH.

8) After one base part and tray have passed the Bit Shift Left function puts a 1 in the most right B3:/0 register. The register will now look like:

Offset	/5	/4	/3	/2	/1	/0
B3:0	0	0	0	0	0	1

9) After two base parts and trays have passed the Bit Shift Left function puts a 1 in the most right register. The register will now look like:

Offset	/5	/4	/3	/2	/1	/0
B3:0	0	0	0	0	1	1

10) New base part and tray combinations will shift the data to the left until all registers are 1, as shown below.

Offset	/5	/4	/3	/2	/1	/0
B3:0	1	1	1	1	1	1

11) Now that B3:0/5 is 1 the contact named CLOSES WHEN B3:0/5 IS 1 closes and activates the output named MATING PART PLACING MACHINE. This places a mating part on the base part.

12) As new trays pass by the 1's of the register will be moved to the left. As long as all trays have base parts all 1's will be shifted to the left, B3:0/5 will be 1 and mating parts will be put on base parts.

13) If an empty tray arrives at the limit switch the register will now look like:

Offset	/5	/4	/3	/2	/1	/0
B3:0	1	1	1	1	1	0

14) If the next tray contains a base part the register will now look like:

Offset	/5	/4	/3	/2	/1	/0
B3:0	1	1	1	1	0	1

15) If the next four trays contain base parts the bits will continue to shift to the left until the register looks like:

Offset	/5	/4	/3	/2	/1	/0
B3:0	0	1	1	1	1	1

16) When B3:0/5 equals to 0 a mating part will not be put on the tray.

5.10.1 Ladder diagram.

Figure 5-28 Ladder diagram for the conveyor belt part placer.

5.10.2 MicroLogix 1000 circuit.

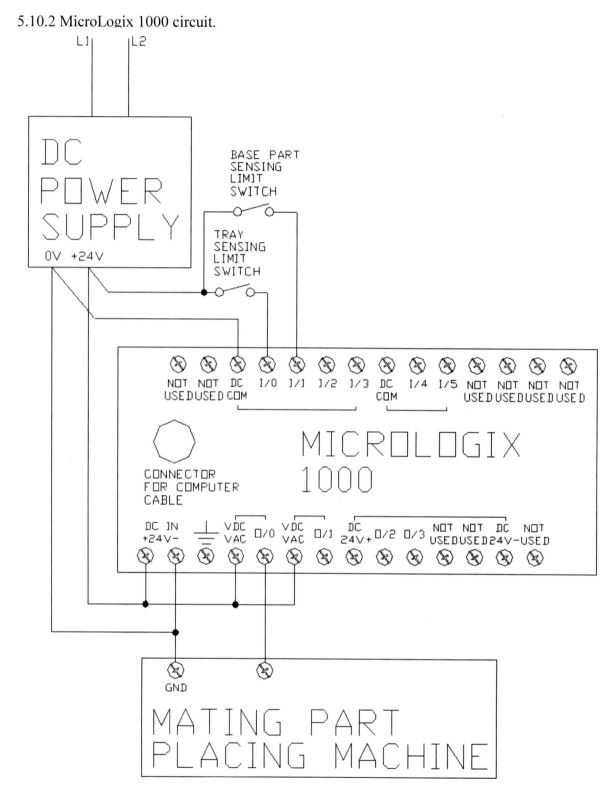

Figure 5-29 MicroLogix 1000 circuit for the conveyor belt part placer.

5.11 MICROLOGIX 1000 LADDER DIAGRAMS AND POWER DIAGRAMS FOR A CONVEYOR BELT SPEED CONTROLLER

In this example a single pushbutton is used to start and set the speed of a conveyor belt. A "Sequencer Output" function is used as the controller, much like a mechanical drum switch. The first pressing of the pushbutton starts the conveyor belt and runs it at low speed. Pressing the pushbutton again speeds up the conveyor belt to medium speed. Pressing it again speeds the conveyor belt to its high speed. Pressing it again stops the conveyor belt. At any speed, pressing the STOP pushbutton resets the conveyor belt system to the stopped condition.

The new material demonstrated here is the use of:
1) The "Sequencer Output" function.
2) The order of ladder program scanning.
3) Binary data, B3:_/_ .

Operation sequence with the ladder diagram of Figure 5-30:
1) The conveyor belt is stopped.
2) Power goes to the MicroLogix 1000.
3) At this time the SEQUENCER OUTPUT function is at the rest position. This is position 1 in the SEQUENCER OUTPUT function. The binary data file for position 1 is shown below and in Figure 5-28.

Offset	/3	/2	/1	/0
B3:0	0	0	0	0
B3:1	0	0	0	1
B3:2	0	0	1	0
B3:3	0	1	0	0
B3:4	1	0	0	0
B3:5	1	1	1	1
B3:6	0	0	0	1

At position 1 the data of the word B3:1 is filtered through the word B3:5 and placed in the word B3:6. Since B3:5 has all 1's each of the B3:1 data goes to B3:6. If there had been 0's in B3:5 the offsets with the 0's would not pass the data of B3:1. For example, if the bit B3:5/0 had been 0 then the bit B3:6/0 would be 0, not 1.

4) The information of B3:6 is used in the lower portion of the ladder diagram to operate MicroLogix 1000 outputs. At position 1, B3:6/0 is 1 so the B3:6/0 "Examine a bit for an ON" goes on and the output O:0/0, to the STOP on the inverter, is on.

5) The SPEED PUSHBUTTON is pushed and released once.

6) The SEQUENCER OUTPUT function receives one off-on-off impulse. This advances the SEQUENCER OUTPUT function to position 2. The binary data file for position 2 is shown below.

Offset	/3	/2	/1	/0
B3:0	0	0	0	0
B3:1	0	0	0	1
B3:2	0	0	1	0
B3:3	0	1	0	0
B3:4	1	0	0	0
B3:5	1	1	1	1
B3:6	0	0	1	0

At position 2 the data of the word B3:2 is filtered through the word B3:5 and placed in the word B3:6. Since B3:5 has all 1's each of the B3:2 data goes to B3:6. If there had been 0's in B3:5 the offsets with the 0's would not pass the data of B3:2.

7) The information of B3:6 is used in the lower portion of the ladder diagram to operate MicroLogix 1000 outputs. At position 2, B3:6/1 is 1 so the B3:6/1 "Examine a bit for an ON" goes on and the output O:0/1, to the LOW SPEED on the inverter, is on.

8) The SPEED PUSHBUTTON is pushed and released once.

9) The SEQUENCER OUTPUT function receives one off-on-off impulse. This advances the SEQUENCER OUTPUT function to position 3. The binary data file for position 3 is shown below.

Offset	/3	/2	/1	/0
B3:0	0	0	0	0
B3:1	0	0	0	1
B3:2	0	0	1	0
B3:3	0	1	0	0
B3:4	1	0	0	0
B3:5	1	1	1	1
B3:6	0	1	0	0

At position 3 the data of the word B3:2 is filtered through the word B3:5 and placed in the word B3:6. Since B3:5 has all 1's each of the B3:2 data goes to B3:6. If there had been 0's in B3:5 the offsets with the 0's would not pass the data of B3:2.

10) The information of B3:6 is used in the lower portion of the ladder diagram to operate MicroLogix 1000 outputs. At position 3, B3:6/2 is 1 so the B3:6/2 "Examine a bit for an ON" goes on and the output O:0/2, to the MEDIUM SPEED on the inverter, is on.

11) The SPEED PUSHBUTTON is pushed and released once.

12) The SEQUENCER OUTPUT function receives one off-on-off impulse. This advances the SEQUENCER OUTPUT function to position 4. The binary data file for position 4 is shown below.

Offset	/3	/2	/1	/0
B3:0	0	0	0	0
B3:1	0	0	0	1
B3:2	0	0	1	0
B3:3	0	1	0	0
B3:4	1	0	0	0
B3:5	1	1	1	1
B3:6	1	0	0	0

At position 4 the data of the word B3:2 is filtered through the word B3:5 and placed in the word B3:6. Since B3:5 has all 1's each of the B3:2 data goes to B3:6. If there had been 0's in B3:5 the offsets with the 0's would not pass the data of B3:2.

13) The information of B3:6 is used in the lower portion of the ladder diagram to operate MicroLogix 1000 outputs. At position 4, B3:6/3 is 1 so the B3:6/3 "Examine a bit for an ON" goes on and the output O:0/3, to the HIGH SPEED on the inverter, is on.

14) The SPEED PUSHBUTTON is pushed and released once.

15) The SEQUENCER OUTPUT function receives one off-on-off impulse. This sends the SEQUENCER OUTPUT function back to position 1. The binary data file for position 1 is again shown below.

Offset	/3	/2	/1	/0
B3:0	0	0	0	0
B3:1	0	0	0	1
B3:2	0	0	1	0
B3:3	0	1	0	0
B3:4	1	0	0	0
B3:5	1	1	1	1
B3:6	0	0	0	1

16) The SPEED PUSHBUTTON and speed cycle of 3) to 15) reoccurs as long as the SPEED PUSHBUTTON is pushed.

17) At any speed, if the STOP pushbutton is pushed the MicroLogix 1000 will stop the conveyor belt.

18) If the conveyor belt is not already stopped and the STOP pushbutton is pushed, the ladder program first holds itself on with the O:0/6, "Examine a bit for an ON". The use of O:0/6 assures that the STOP function will remain active even if the STOP pushbutton was only very briefly opened.

19) O:0/6 being ON starts a sequence of operations in the ladder program.

20) A significant difference between relay ladder diagrams and programmable ladder diagrams is that the rungs on a relay ladder are energized simultaneously, but the rungs on programmable controller ladder diagram are enabled one at a time. The rungs on a programmable controller are scanned through. With Allen-Bradley programmable controllers the rungs are read and executed left to right across the rungs and then down to the left side of the next down rung in a zigzag pattern.

21) The scanning of the rungs will cause O:0/4 and O:0/5 to turn each other on and off. In the process the input to the SEQUENCER OUTPUT will also be turned on and off. The turning on and off will continue until the STOP output, O:0/0, is on. Then the ladder program will rest in the STOP position.

5.11.1 Ladder diagram for the conveyor belt speed controller with the Sequencer Output connected to intermediate Binary functions.

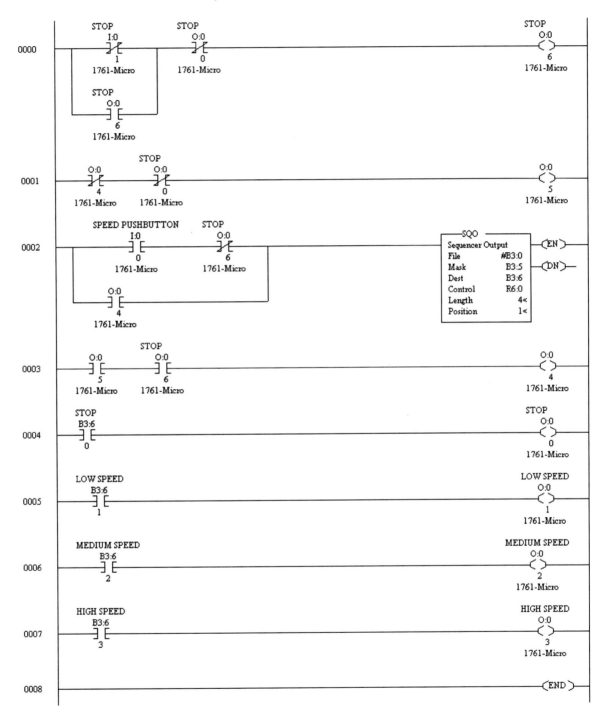

Figure 5-30 Ladder diagram for the conveyor belt speed controller with the Sequencer Output connected to intermediate Binary functions.

5.11.2 Binary Data File

Binary data needs to be entered into the RSLogix program in addition to the ladder diagram. Figure 5-31 shows the Binary Data File window that can be called up from RSLogix under the Project Folder for the ladder diagram of Figure 5-30. Under the Project Folder it is in the Data Files Folder in B3-Binary.

Once the window of Figure 5-31 has been called up appropriate 1's and 0's are entered. For example, to enter a 1 at B3:1/0 first put the cursor over the location. Left click the mouse to highlight the location. Then type 1 and press Enter. 1 should now be added that location. Place 1's and 0's as shown in Figure 5-28.

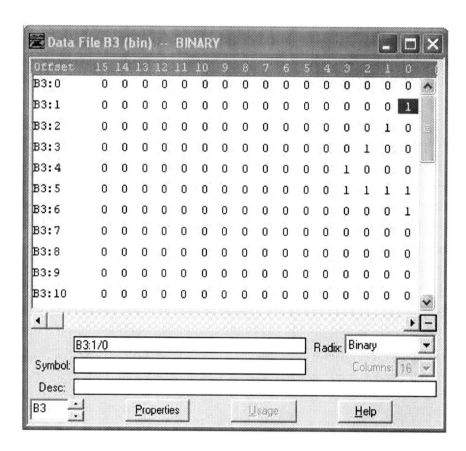

Figure 5-31 Binary Data File for the conveyor belt speed controller.

5.11.3 Ladder diagram for the conveyor belt speed controller with the Sequencer Output connected directly to outputs

It is possible to have the SEQUENCER OUTPUT directly control the MicroLogix 1000 outputs without using the binary functions that were used in Figure 5-30. This has been done in Figure 5-32. Probably an experienced programmer would prefer this method, if the other outputs in O:0 were not needed elsewhere in the ladder.

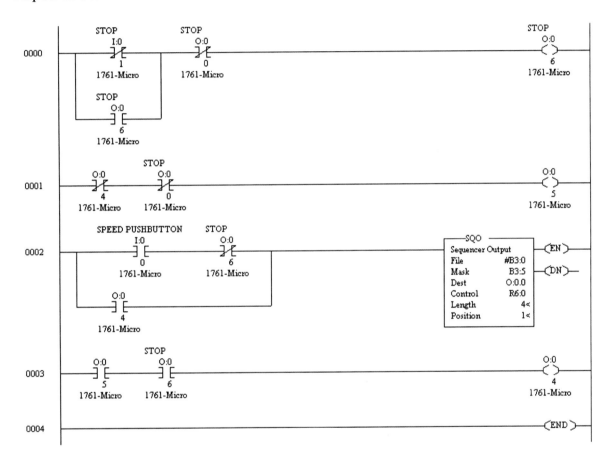

Figure 5-32 Ladder diagram for the conveyor belt speed controller with the Sequencer Output connected directly to Outputs.

72

5.11.4 MicroLogix 1000 circuit.

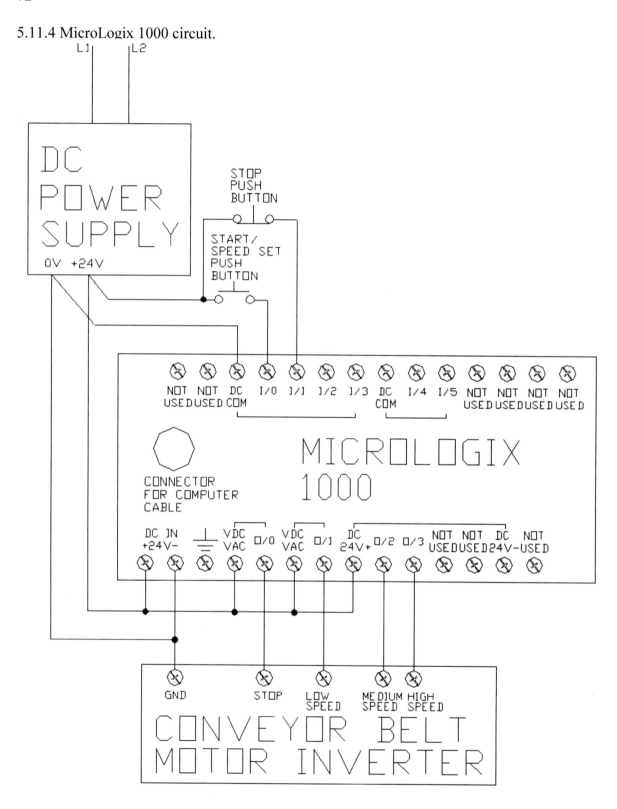

Figure 5-33 MicroLogix 1000 circuit for the conveyor belt speed controller.

5.11.5 Power circuit.

The same power circuit is used here as was used in Section 5.9. See Figure 5-27.

5.12 MICROLOGIX 1000 LADDER DIAGRAM AND POWER DIAGRAM FOR A SYSTEM THAT USES FIFO DATA STACK STORAGE FOR SELECTING SPRAY PAINT COLORS

In this example the MicroLogix 1000 is used to control paint spray color in an assembly line painting booth. When assembly line items enter the holding area prior to the paint booth a bar code reader reads the paint colors from bar codes attached to the items. The bar code reader closes the appropriate color switch for each item. The MicroLogix 1000 stores the color selection in a stack of color codes. When an item leaves the holding area, the MicroLogix 1000 uses the code for the oldest entered color code to select the appropriate color spray paint. It then forgets that color code. This selection and one time using of the oldest color code is done with the First In First Out (FIFO) function.

The new material demonstrated here is the use of:
1) The "FIFO Load" function.
2) The "FIFO Unload" function.
3) Integers, N7:_, as words in the FIFO stack..

Operation sequence with the ladder diagram of Figure 5-34:
1) The assembly line is stopped and the paint booth entry holding area is empty.
2) Power goes to the MicroLogix 1000.
3) The assembly line starts and sends an item to the paint booth entry holding area.
4) A bar code reader at the entrance to the holding area reads the item's bar code, interprets the code, and closes the appropriate switch connected to the MicroLogix 1000 input.
5) The MicroLogix 1000 stores the color code as the first received in a data file stack. The colors are stored in the MicroLogix 1000 with the following words: Red = N7:4/0 = 2^0 decimal = 1 decimal, Yellow = N7:4/1 = 2^1 decimal = 2 decimal, and Green = N7:4/3 = 2^3 decimal = 8 decimal.
6) In this example suppose the entered item needs to be painted red, then the Source, N7:4 will be set to 1.
7) The FFL (FIFO Load) places value of N7:4, 1, in the 4 long data stack at position N7:0/0. The stack now looks like:

Offset	/3	/2	/1	/0
N7:0	0	0	0	1

8) The output word N7:10 is 0.
9) Suppose the next entered item needs to be painted yellow, then the data stack will be changed to:

Offset	/3	/2	/1	/0
N7:0	0	0	2	1

10) Suppose the next entered item needs to be painted green, then the data stack will be changed to:

Offset	/3	/2	/1	/0
N7:0	0	8	2	1

11) Suppose the next entered item needs to be painted red, then the data stack will be changed to:

Offset	/3	/2	/1	/0
N7:0	1	8	2	1

12) Notice how this looks like the BSR function (Bit Shift Right). However, here words have been shifted rather than bits. Also, here 0's will not be shifted, but will be written over.

13) The output word, N7:10 has been 0 throughout this process, since the items have not operated the Spray Booth Entry Limit Switch and caused the FFU to unload the stack.

14) When the FFU does operate the first time it will send the red word, 1, to the EQU functions causing the red spray paint solenoid valve to receive power and go on.

15) At the same time the words will shift to the right so that the data stack will be changed to:

Offset	/3	/2	/1	/0
N7:0	0	1	8	2

16) Now that the N7:0/3 has been emptied it is available for new data. Another item could be put in the holding area and its color word entered into N7:0/3.

76

5.12.1 Ladder diagram.

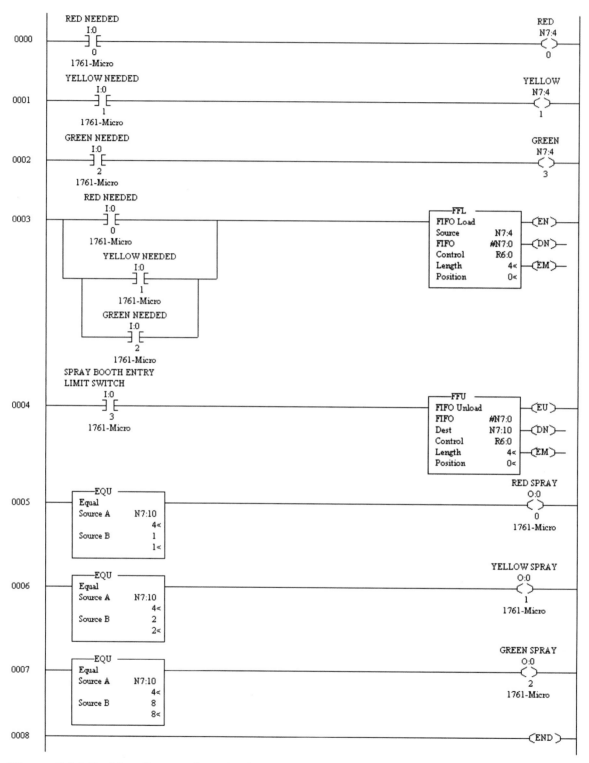

Figure 5-34 Ladder diagram for selecting spray paint colors.

5.11.4 MicroLogix 1000 circuit.

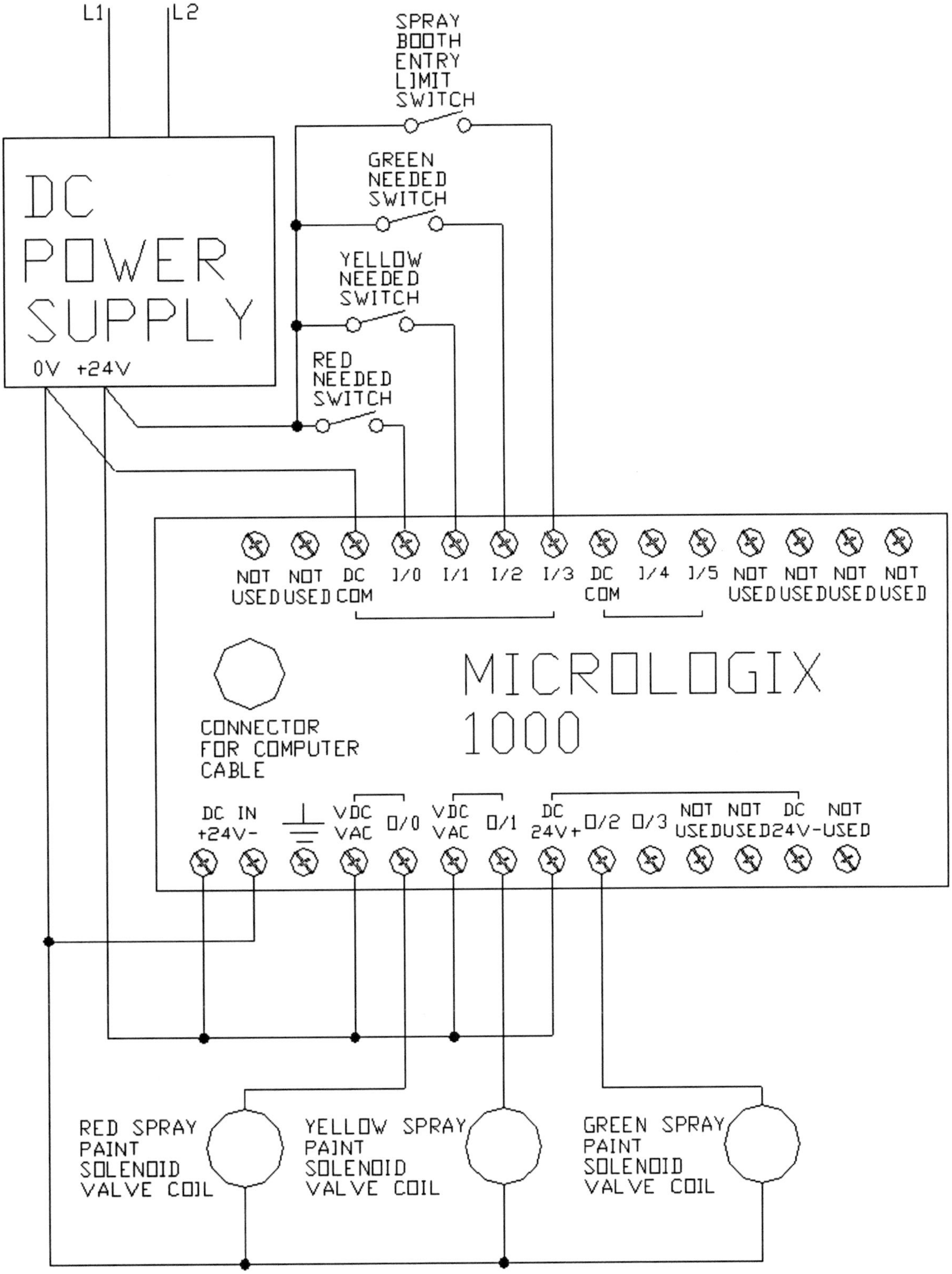

Figure 5-35 MicroLogix 1000 circuit for selecting spray paint colors.

5.13 MICROLOGIX 1000 LADDER DIAGRAM AND POWER DIAGRAMS FOR A SYSTEM THAT USES HIGH SPEED COUNTING TO MEASURE AND CONTROL MOTOR SPEED

In this example the MicroLogix 1000 high speed counter is used to measure the forward and reverse rotation increments (counts) produced by an induction motor driven encoder over 15 second time periods. The number of counts is used to set the output frequency of the inverter that drives the motor.

The MicroLogix 1000 Count Up and Count Down counters can only count one count per program scan. The high speed counter counts independently of the program scan, allowing it to count faster. The MicroLogix 1000 high speed counter can handle counting input frequencies of up to 6.6 kHz. If an encoder produced 100 counts per revolution then motor speeds could be measured up to (6600/100)x60 = 3960 rpm.

Only one high speed counter can be used per MicroLogix 1000. When the high speed counter is configured for an encoder the counting inputs come in through I:0/0 and I:0/1. The order in which the counts come in tell the MicroLogix 1000 whether the encoder is going forward or reverse. Forward produces positive counts. Reverse produces negative counts. The high speed counter can also be configured in other ways that may use some or all of the inputs I:0/0, I:0/1, I:0/2, and I:0/3.

The new material demonstrated here is the use of:
1) The "High Speed Counter" function.
2) The "HSC Reset Accumulator" function.

Operation sequence with the ladder diagram of Figure 5-36:
1) Power goes to the MicroLogix 1000.
2) The High Speed Counter starts counting and the Timer On Delay starts timing.
3) After 15 seconds the Limit Test functions compare the count to two limit ranges. If the count is from 0 to 3 then output O:0/0 will be enabled. If the count is from 4 to 7 then O:0/1 will be enabled. The limits here are low so that they can be met with the manually operated switches in the teaching setup. The High Speed counter is capable of counting up to a maximum of 32,767 counts in 5 seconds.
4) The MicroLogix 1000 will determine the direction of rotation by the order of the turning on and off of the inputs I:0/0 and I:0/1. If the order of turning on and off is I:0/0 on, I:0/1 on, I:0/0 off, I:0/1 off, I:0/0 on, I:0/1 on and so on, the count number will be positive. If the order of turning on and off is I:0/1 on, I:0/0 on, I:0/1 off, I:0/0 off, I:0/1 on, I:0/0 on and so on, the count number will be negative. This can be verified with the teaching setup.
5) After the ladder program turns on the appropriate output, the program resets the high speed counter and timer to zeros.
6) The counting and timing starts again.

5.13.1 Ladder diagram.

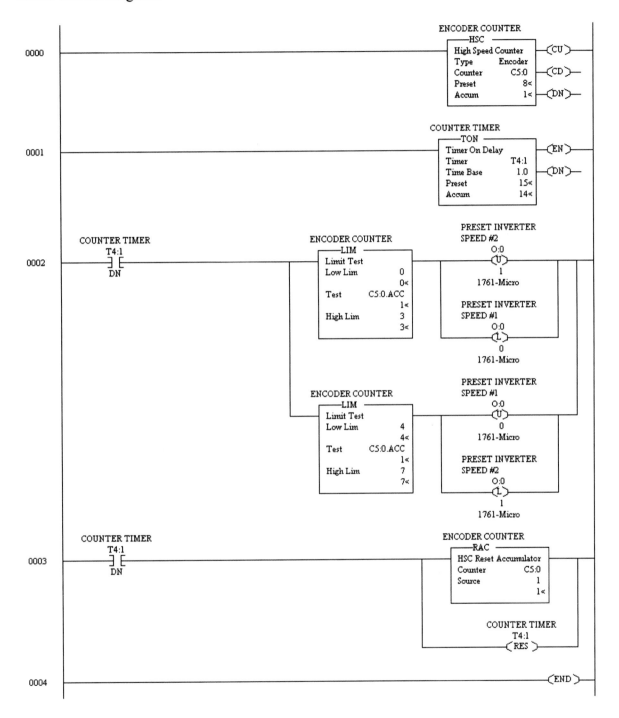

Figure 5-36 Ladder diagram for the encoder control circuit.

80

5.13.2 MicroLogix 1000 circuit.

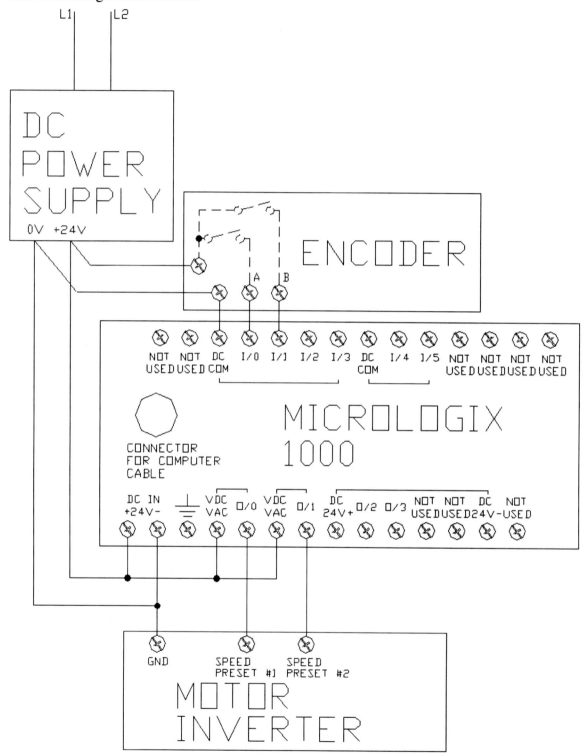

Figure 5-37 MicroLogix 1000 circuit for the encoder control circuit.

5.13.3 Power circuit.

The same power circuit is used here as was used in Section 5.9. See Figure 5-27.

6.0 OTHER SOURCES OF INFORMATION

6.1 USEFUL REFERENCE BOOKS

Cox, R. A., 2001. *Technician's Guide to Programmable Controllers,* 4th Ed. Albany, NY: Delmar Thomson Learning, price $95.95.
This is an excellent book. Anybody investing the time in learning about PLCs should get a copy.

Rockwell Automation Allen-Bradley, 1995. *MicroMentor Understanding and Applying Micro Programmable Controllers,* Catalog Number 1761-MMB, 1201 South Second Street, Milwaukee, WI 53204 tel 414-382-4444: Allen-Bradley Headquarters, price $20.00.
The descriptions and examples given in this book are for the MicroLogix PLCs and RSLogix 500 software. It is a good book to go to for a second explanation of topics covered in Cox's book. Purchase it directly from Allen-Bradley.

To get free downloadable MicroLogix Information:
Go to the web address http://www.ab.com/plclogic/micrologix/.
Look under 'Get Information' and select from the following:
'MicroLogix Brochure 1761-BR00A-EN-P' A general brochure.
'MicroLogix Selection Chart' A chart for selecting the correct MicroLogix PLC.
'MicroLogix Technical Datas – MicroLogix 1000'. This refers to a number of MicroLogix information downloads. Among these are:
Installation Instructions, 1761-5.1.2
User Manual, 1761-6.3
Technical Data, 1761-TD001B-EN-P

To get free downloadable RSLogix 500 information:
1) Go to:
http://literature.rockwellautomation.com/idc/groups/public/documents/webassets/brow se_category.hcst
2) Search for LG500-GR001A-EN-P.
3) This will take you to a page where you can download *RSLogix500 Getting Results Guide*, DOC ID LG500-GR001A-EN-P.

To get free downloadable RSLinx information:
   1) Go to:
   http://literature.rockwellautomation.com/idc/groups/public/documents/webassets/browse
   _category.hcst
   2) Search for *Getting Results with RSLinx*.
   3) This will take you to a page where you can find and download *Getting Results with RSLinx,* Pub. No. 9399-WAB32GRJA-JUN98.

6.2 ALLEN-BRADLEY CONTACTS

Allen-Bradley, division
of Rockwell Automation
1201 South Second Street
Milwaukee, WI 53204
1-414-382-2000
http://www.ab.com/

Allen-Bradley has courses on the use of their PLCs and other products. The courses are very good, but may be too expensive for individuals. Most of the students in these courses have their tuition paid by their employers.

Local Allen-Bradley sales offices will sometimes provide free help to those using or teaching themselves how to use Allen-Bradley PLCs. Sales office locations near to you can be provided by the corporate Allen-Bradley office.

6.3 ELECTRICAL SUPPLIERS THAT SELL ALLEN-BRADLEY PLCS

Electrical supply companies sometimes offer free instructional/sales seminars on Allen-Bradley PLCs. If you purchase an inexpensive MicroLogix 1000 from one of these companies it would be easier to later call them for technical help.

7.0 APPENDIX

7.1 RELAY LADDER AND POWER DIAGRAM SYMBOLS

Relay or contactor coil

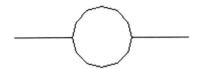

Normally open relay or contactor contacts

Normally closed relay or contactor contracts

Normally open pushbutton

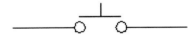

Normally closed pushbutton

Normally open single pole switch or disconnect

Normally closed single pole switch

Circuit breaker

Pilot light

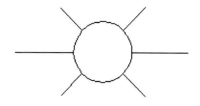

Three-phase induction motor

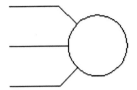

7.2 RSLOGIX 500 LADDER DIAGRAM FUNCTION SYMBOLS

These commonly used function symbols have been taken from the examples of chapter 5.0. There are many other useful functions. Information on all functions can be seen in the RSLogix 500 program by clicking "Help" and under that "SLC Instruction Help".

BASIC USER SYMBOLS

Examine a bit for an ON condition
This is equivalent to a relay ladder diagram's open relay contacts. I:0 and 0 identify the input connection point to the MicroLogix 1000. When the input I:0/0 is connected to +24 VDC this symbol acts like a closed relay in a relay ladder diagram. START is an optional name. See any of the example ladder diagrams from Figures 5-5 to 5-34.

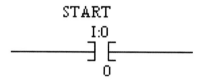

Examine a bit for an ON condition
This is equivalent to a relay ladder diagram's open relay contacts. O:0 and 0 identify the output connection point to the MicroLogix 1000 or contacts used elsewhere in the ladder diagram. When the output O:0/0 is enabled this symbol acts like a closed relay in a relay ladder diagram. FORWARD CONTACTOR is an optional name. See Figures 5-5, 5-8, 5-9, 5-12, 5-14, 5-19, 5-21, 5-30, and 5-32.

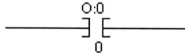

Examine a bit for an ON condition

This is equivalent to a relay ladder diagram's open relay contacts. T4:0 is its name and that of its corresponding Timer. DN shows that the contacts will close like a closed relay in a relay ladder diagram when the Timer has timed to its preset time limit. See Figures 5-9, 5-21, 5-23, 5-25, and 5-36.

```
                 T4:0
  ───────────────┤ ├───────────────
                  DN
```

Examine a bit for an ON condition

C5:0 is its name and the name of its corresponding up counter and/or down counter. DN shows that the contacts will close like a closed relay in a relay ladder diagram when the Counter has counted to its preset limit. See Figures 5-12, 5-14, and 5-21.

```
                 C5:0
  ───────────────┤ ├───────────────
                  DN
```

Examine a bit for an ON condition

This does not have an equivalent in an ordinary relay ladder diagram. B3:6/2 is its name. When bit B3:6/2 is 1 it is enabled and acts like closed contacts in a relay ladder diagram. When it equals 0 it is not enabled. MEDIUM SPEED is an optional name. See Figures 5-28 and 5-30.

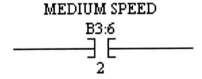

```
        MEDIUM SPEED
             B3:6
  ───────────┤ ├───────────
              2
```

90

Examine a bit for an OFF condition

This is equivalent to a relay ladder diagram's normally closed relay contacts. O:0 and 2 identify the output connection point to the MicroLogix 1000 or contacts used elsewhere in the ladder diagram. When the output O:0/2 is enabled this symbol acts like an open relay in a relay ladder diagram. Any of the "Examine for a bit for an ON condition" locations (i.e. I:0, T4:0, C5:0, etc.) or conditions (i.e. 0, 2, DN, etc.) will work with this. STOP is an optional name. See Figures 5-5, 5-8, 5-9, 5-12, 5-14, 5-19, 5-21, 5-30, and 5-32.

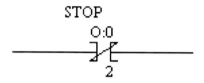

Turn ON a bit, non-retentive

This is equivalent to a relay diagram's relay coil. O:0 and 0 identify the output connection point or the MicroLogix 1000 or contacts used elsewhere in the ladder diagram. FORWARD CONTACTOR is an optional name. See Figures 5-5, 5-8, 5-9, 5-12, 5-14, 5-16, 5-19, 5-21, 5-28, 5-30, 5-32, and 5-34.

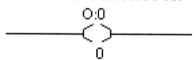

Turn ON an integer, non-retentive

This does not have an equivalent in an ordinary relay ladder diagram. N:7 and 4 identify the integer location, 3 is exponent of 2 of the integer (integer $= 2^3 = 8$). GREEN is an optional name. See Figure 5-34.

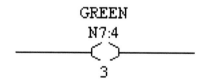

Select a bit ON, retentive and latching

This is equivalent to a relay ladder diagram's latching relay's latching coil. Once briefly enabled its corresponding contacts will remain enabled until a corresponding unlatching symbol is briefly enabled. O:0 and 2 identify the output connection point of the MicroLogix 1000. STOP is an optional name. See Figures 5-9, 5-19, 5-23, 5-25, 5-36.

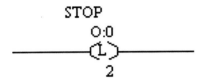

Select a bit OFF, retentive and unlatching

This is equivalent to a relay ladder diagram's latching relay's unlatching coil. Once briefly enabled its corresponding latching contacts will be un-enabled. O:0 and 2 identify the output connection point of the MicroLogix 1000. STOP is an optional name. See Figures 5-9, 5-19, 5-23, 5-25, 5-36.

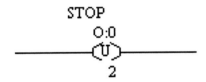

TIMER AND COUNTER SYMBOLS

Delay turn on a bit

This is equivalent to a relay ladder diagram's time delay relay coil. When it is enabled it starts timing. Once the timer has reached its preset value it operates its corresponding contacts. T4:0 is its name and the name of its corresponding contacts. The Time Base, 1.0, shows that it will time in 1 second intervals. This time base could be set at smaller values, .01 seconds, for example. The Preset, 3<, indicates the timer will operate in 3 time intervals, which is 3 seconds in this case. The Accum, 0, indicates the number of time intervals that have passed. See Figure 5-9, 5-19, 5-23, 5-25, and 5-36.

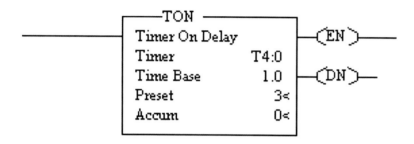

Count Up

This does not have an equivalent in an ordinary relay ladder diagram. Each time it is enabled it advances its C5:0 counter one count. C5:0 is the identification number of this Count Up counter, perhaps a corresponding Count Down counter, and their corresponding contacts. Preset indicates the number that it will count up to before operating, 6< in this case. The Accum indicates the number it has counted up to so far, 0< in this case. See Figures 5-12, 5-14, 5-16, 5-21, 5-23, and 5-25.

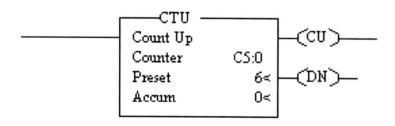

Count Down

This does not have an equivalent in an ordinary relay ladder diagram. Each time it is enabled it decreases its C5:0 counter one count. C5:0 is the identification number of this Count Down counter, perhaps a corresponding Count Up counter, and their corresponding contacts. Preset indicates the number that it will count down to before operating, 6< in this case. The Accum indicates the number it has counted down to so far, 0< in this case. See Figure 5-16.

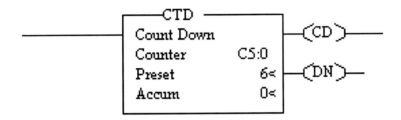

Reset accumulated value and status bits of a Timer

T4:1 is the identification number of the corresponding Timer. When this is enabled the Accum value in the Timer T4:1 is reset to zero. See Figures 5-23, 5-25, and 5-36.

```
        T4:1
————————( RES )————————
```

Reset accumulated value and status bits of a Counter

C5:0 is the identification number of the corresponding Count Down or count Up counter. When this is enabled the Accum value is reset to zero in the C5:0 Count Up and/or Count Down counters. See Figures 5-12, 5-14, 5-21, 5-23, and 5-25.

```
        C5:0
————————( RES )————————
```

COMPARE SYMBOLS

Limit Test

This does not have an equivalent in an ordinary relay ladder diagram. When the quantity in Test is between the Low Lim and High Lim the block acts like closed contacts in a relay ladder diagram. In this case the Test value is N7:0 = 2, the Low Lim is 2 and the High Lim is 4 so the block acts like closed contacts since N7:0 equals 2. See Figure 5-25 and 5-36.

```
            ┌───LIM ────────┐
────────────┤ Limit Test    ├────────────
            │ Low Lim      2 │
            │             2< │
            │ Test      N7:0 │
            │             2< │
            │ High Lim     4 │
            │             4< │
            └───────────────┘
```

Equal

This does not have an equivalent in an ordinary relay ladder diagram. When the quantity in Source A, the time value in T4:1.ACC in this case, is equal to the quantity in Source B, 15 in this case, the EQU block acts like closed contacts in a relay ladder diagram. See Figure 5-23 and 5-34.

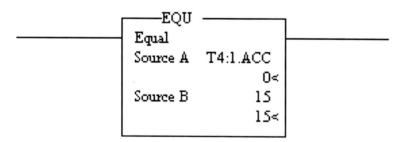

```
          ┌───EQU ────────────┐
──────────┤ Equal             ├──────────
          │ Source A  T4:1.ACC │
          │               0<   │
          │ Source B     15    │
          │              15<   │
          └────────────────────┘
```

Greater Than

This does not have an equivalent in an ordinary relay ladder diagram. When the quantity in Source A, the count in C5:0.ACC in this case, is greater than the quantity in Source B, N7:1 = 4 in this case, the GRT block acts like closed contacts in a relay ladder diagram. See Figure 5-23.

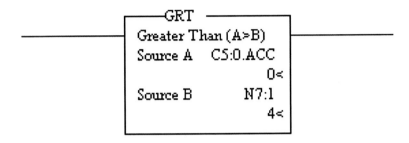

Less Than

This does not have an equivalent in an ordinary relay ladder diagram. When the quantity in Source A, the count in C5:0.ACC in this case, is less than the quantity in Source B, N7:2 = 2 in this case, the LES block acts like closed contacts in a relay ladder diagram. See Figure 5-23.

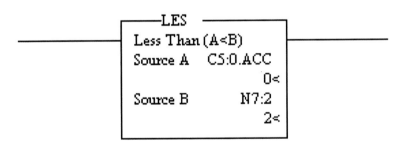

COMPUTATION SYMBOL

Addition

This does not have an equivalent in an ordinary relay ladder diagram. The quantity in Source A is added to the quantity in Source B and the result placed in the space in Dest. In this case the quantity in C5:0.ACC is added to 2 and the result is placed in N7:0. Here that is $0 + 2 = 2$. See Figure 5-25.

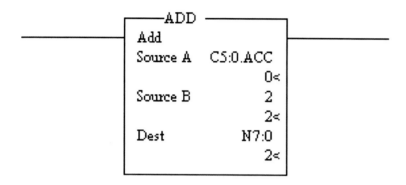

MOVE SYMBOL

Transfer contents of source to destination

This does not have an equivalent in an ordinary relay ladder diagram. Each time it is enabled it moves data from the Source to the Dest. Here the constant 8 is moved to the Preset value of counter C5:0. See Figure 5-14.

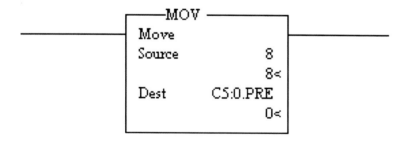

FILE SHIFT AND SEQUENCER SYMBOLS

Bit Shift Left

This does not have an equivalent in an ordinary relay ladder diagram. The File contains the address of the bit file that will contain the 0 and 1 bit data. In this case the bit data is in B3:0. The # must precede the B3:0. The Control states where the control structure for the BSL is stored. In this case it is in R6:1. The Bit Address is the location of the bit input. In this case the bit input comes from input I:0/1. The Length is a statement of how many bits are to be shifted to the left. In this case there are 6 bits to shift. Here these are B3:0/0, B3:0/1, B3:0/2, B3:0/3, B3:0/4, and B3:0/5. See Figure 5-28.

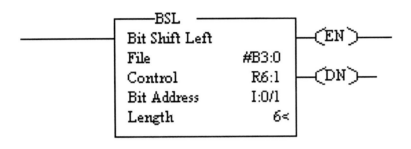

98

Sequencer Output

This does not have an equivalent in an ordinary relay ladder diagram, although its output and use are similar to a drum controller's. The File contains the address of the bit file that will contain the bit data. In this case the bit data is in B3:0. # must precede the B3:0. In this case File data is stored in words B3:1, B3:2, B3:3, and B3:4. The Mask states the location of the filter through which File data must pass. 1 bits in the Mask will cause bits in the File data to be rewritten in the Dest address. 0 bits in the Mask will cause bits in the File data to be ignored in the Dest address. In this case the Mask data is stored in word B3:5. The Dest address receives Mask filtered File data from the File. In this case bits go to outputs B3:6/0, B3:6/1, B3:6/2, and B3:6/3. The Control gives the location of SQO's control address. The Length give the number of sequencer steps past word in the File. In this case the Length is 4, meaning there are 4 words after B3:0. The Position indicates the word that the File is currently sending through the Mask to the Dest. In this case the Position is 1, indicating that the File is sending data from word B3:1. See Figures 5-30, 5-31, and 5-32.

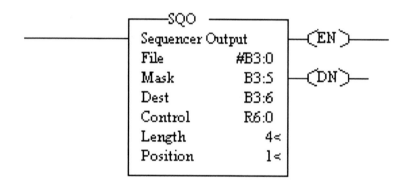

FIFO Load

This does not have an equivalent in an ordinary relay ladder diagram. The FIFO Load is used in combination with the FIFO Unload. The Source is the location of the integer word that is next loaded into the FIFO (First In First Out) data stack. In this case it is N7:4. FIFO states the starting address of the data stack. The # symbol must appear to the left of the address. In this case it starts at word N7:0. Control is the control address, R6:0 in this case. Length is the maximum number of words in the stack. In this case it is 4. Since the stack starts at N7:0 the stack contains the 4 words N7:0/0, N7:0/1, N7:0/2, and N7:0/3. Position is the next available location where a Source word would be loaded. In this case position 0, N7:0/0, is the next available location. See Figure 5-34.

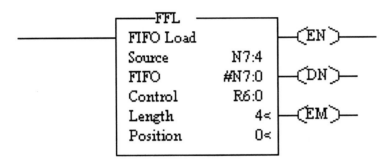

FIFO Unload

This does not have an equivalent in an ordinary relay ladder diagram. The FIFO Unload is used in combination with the FIFO Load. FIFO states the starting address of the data stack. The # symbol must appear to the left of the address. In this case it starts at word N7:0. Dest states the address the data taken from the stack is to be sent to. In this case it is N7:10. Control is the control address, R6:0 in this case. Length is the maximum number of words in the stack. In this case it is 4. Since the stack starts at N7:0 the stack contains the 4 words N7:0/0, N7:0/1. N7:0/2, and N7:0/3. Position is the next available location where a Source word would be loaded. In this case position 0, N7:0/0, is the next available location. See Figure 5-34.

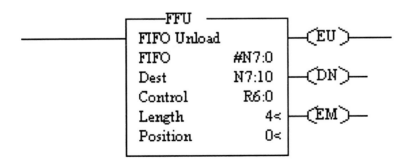

MICROLOGIX 1000 HIGH SPEED COUNTER

High Speed Counter

This does not have an equivalent in an ordinary relay ladder diagram. The item in Type is the type of high speed counter. There are eight types of high speed counters. In this case the high speed counter selected is the Encoder type. It is configured to receive input counts from I:0/0 and I:0/1. The high speed counter will determine the direction of rotation of an encoder by the order of the turning on and off of the inputs I:0/0 and I:0/1. If the order of turning on and off is I:0/0 on, I:0/1 on, I:0/0 off, I:0/1 off, I:0/0 on, I:0/1 on and so on, the count number will be positive. If the order of turning on and off is I:0/1 on, I:0/0 on, I:0/1 off, I:0/0 off, I:0/1 on, I:0/0 on and so on, the count number will be negative.

The address in Counter is C5:0. That location is required by the MicroLogix 1000 high speed counter and can not be used elsewhere when the high speed counter is used.

The Preset is the value that the high speed counter will count to before it automatically updates outputs or generates a high speed counter interrupt.

The Accum has the value the high speed counter has counted to so far. In this case the 1 shows that 1 count has been recorded.

Only one high speed counter is available on a MicroLogix 1000. The MicroLogix 1000 high speed counter is different from that of all other Allen-Bradley high speed counters. See Figure 5-36.

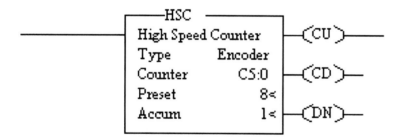

HSC Reset Accumulator

This does not have an equivalent in an ordinary relay ladder diagram. This resets the high speed counter to 0. Counter refers to address C5:0. This is always the same address with the HSC Reset Accumulator. Source refers to a constant that is loaded into the accumulator. In this case 1 is the value loaded. See Figure 5-36.

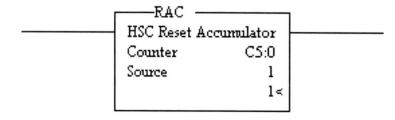

END SYMBOL

Ladder diagram end statement, automatically placed with RSLogix 500
This is at the end of all the RSLogix 500 ladder diagrams.

7.3 PROGRAMMABLE CONTROLLER GLOSSARY

Address: Identifying numbers and/or letters that designate a particular I/O location on a programmable controller, on a device controlled by a programmable controller, or a memory location in a programmable controller.

Adapter Board: Module that is used on external devices to allow them to be connected to a programmable controller's data bus.

Bit: Generally speaking this is a binary digit, a 1 or 0. With programmable controllers using ladder logic this is considered an "on" or "off". An "on" is a continuous path across a ladder rung an "off" is a break in the circuit of a ladder rung.

Cascading: A programming technique of feeding a timer output into a counter input to extend the timing range of the programmable controller beyond that of the timer alone.

Central Processing Unit: Major component of a computer system with the circuitry to control the interpretation and execution of instructions. Often it is abbreviated as CPU. Every programmable controller contains a central processing unit.

Contactor: Are like relays but they switch larger voltages and currents. Their sizes range from that of a book to that of an automobile. Smaller contactors are used to handle motor or lighting loads. The largest contactors are in power company switchyards and can handle thousands of volts and amps. In schematics, they are represented with the same symbols as relays.

EEPROM: Electrically erasable programmable read-only memory.

Enabled: A term to indicate that a function has been activated.

File: A group of consecutive words.

Firmware: Programmable controller instructions that are embedded in the hardware, stored in PROM, EPROM or EEPROM devices, and are generally not modifiable by the user.

Instruction Set or Instructions: These are the program statements used to manipulate data received by the programmable controller. Some of the instructions available on programmable controllers are: relay-type ("on" and "off"), timer, counter, comparison, arithmetic, Boolean logic, and program control.

Interposing Relay: This is a relay placed between one relay or controller output and a higher current or voltage load. For example, a PLC output with a 1.0 amp maximum capability might be needed to control a large contactor that requires a 5.0 amp coil input. To do this the PLC would control an interposing relay whose output contacts can handle the 5.0 amps.

I/O: Abbreviation for "Input/Output". A programmable controller receives data through its input terminals and sends out signals and controlling voltages through its output terminals.

Jog: A state of momentarily being on or in motion. A jog pushbutton allows a machine operator to "inch" a machine forward or reverse as long as the jog pushbutton is pushed. As soon as the jog pushbutton is released the machine stops.

Ladder Diagram: Standard method of drawing relay or logic control circuits. The drawings resemble a ladder. Most programmable controller manufacturers use software created ladder diagrams as part of their programming languages.

Module: Interchangeable plug-in electronic item, often a printed circuit board or card.

Open Architecture: Computer or programmable controller design for which detailed specifications are published by the manufacturer, allowing others to produce compatible hardware and software.

Port: Connector or terminal strip used to access a system or circuit. Usually ports are used for the connection of peripheral equipment.

Programming Terminal: Keyboard device used to input programs and data and operate a programmable controller. It can be mounted on the programmable controller, a separate hand-held device, or a specially configured personal computer.

Rack: Framework or chassis that houses programmable controller modules. From a programming prospective, a single programmable controller framework or chassis sometimes contains more than one rack.

Relay: Electrically operated switch. When voltage is applied to a relay's coil, a magnetic field is produced to move an iron core that mechanically opens or closes electrical contacts. Typical voltage ratings for relay coils are between 12 and 115 volts, AC or DC. Usually, the current rating of a relay's contacts are a few amperes or less.

In many circuit diagrams, circles represent relay coils, and parallel lines represent relay contacts. A relay's coil is designated by the same letters as its contacts. Voltage applied to the A coil will close the A contacts.

There are also more complicated electromagnetic and solid state time-delay, voltage level sensing, current level sensing, and other value sensing relays.

Register: Data storage location in a computer or programmable controller. This could be a word or group of words.

Retentive Register: Data storage location that retains its data during a power down.

Scan Time: The time to read all inputs, execute the control program, and update all input and output statuses. It is the time required to activate an output that is being controlled by a programmable controller. If the scan time is too long, a programmable controller may not be able to successfully control a process.

Statement Language: Sometimes abbreviated as STL. Programming language for programmable controllers that use statements like those found in the BASIC, C++, and other computing languages rather than ladder diagrams.

Upward Migration: Term that indicates that a programmable controller can do everything that its previous model could do, plus some additional functions.

Word: A group of bits in a sequence.

7.4 MICROLOGIX 1000 SPECIFICATIONS

There are 14 models (11 discrete models with 10 to 32 point I/O configurations and 3 analog models with 5 analog I/O and 20 discrete points).

RS-232 port provides network connectivity.

Inputs may be AC, DC and/or analog currents and voltages.

Outputs via relays, TRIACs, MOSFETs, and analog current or voltage.

Both AC and DC powered models.

The smallest models, 10 to 16 I/O, are 120 x 80 x 40 mm (4.72"x3.15"x1.57").

Execution time for a typical 500 instruction program is 1.56 milliseconds, with a throughput of 1.85 milliseconds.

1 Kilobyte memory.

Programmable with software through Windows 95 and newer. Windows 98 or newer required for current RSLogix ladder programming and RSLinx connection software.

Programmable with a Hand Held Programmer.

Application program has more than 735 words.

More than 250 data words available.

Data elements include 512 bits, 40 timers, 32 counters, 16 control files, 105 integer files, and 33 diagnostic states.

Software and documentation available in five languages.

12 basic logic instructions, 43 applied control instructions, and 14 advanced application-specific instructions.